GREAT RAIDS IN HISTORY

Great Raids in History

From Drake to Desert One

Edited by
SAMUEL A. SOUTHWORTH

CASTLE BOOKS

This edition published in 2002 by Castle Books,
A division of Book Sales Inc.
114 Northfield Avenue
Edison, NJ 08837

Reprinted by arrangement with Sarpedon
An imprint of Combined Publishing
476 W. Elm Street
Conshohocken, PA 19428

ISBN 0-7858-1406-X

Contents

Preface and Acknowledgments

When we at Sarpedon first contemplated how to bring the concept of a book on great raids in history to fruition, we had three options. The first was to commission an author with wide-ranging interests and a sure-handed grasp of disparate military periods; the second was to assemble a compilation of previously-published works (after all, certain subjects of this book—Mosby, Custer, Lawrence, Skorzeny—left behind their own accounts, however reliable); or, as a third option, we could seek to create a collaborative effort with new works and fresh perspectives from a variety of current military historians.

Needless to say, we chose the latter option and the result, from this perspective, was the most enjoyable production of a book in the history of this publishing house. Each chapter that came through the door (some sooner than others!) was eagerly anticipated, not only as another building block in the creation of a new book, but because we were anxious to hear the stories ourselves. If readers, in fact, gain as much new information, as well as food for thought, from this book as have the editors, the endeavor will have become more than worthwhile.

Dealing with seventeen different authors on one book was a new experience for us, and one viewed with perhaps a small amount of apprehension as the project began. However, the process was seamless and the skill, expertise and generosity of spirit we encountered in every case on the part of contributing authors was exemplary. Further, for anyone who has not yet witnessed the vibrancy of thought that is currently taking place throughout the military history community, we can only describe this as overwhelming. The scholarship and ongoing analyses by modern writers who are avidly examining, through primary history or historiography, the multitude of dramatic events in our miliary heritage is indeed impressive.

As for the procedure involved in the creation of this book, the topics emanated from the editor and his consultants at 166 Fifth Avenue; the original writing came from authors based everywhere from Fort Leavenworth to York, England. The events to include in a book on raids, however, were continually under intense scrutiny. What, for example, of air power? Nearly every bombing mission is, by definition, a "raid." Many naval actions and, to a significant degree, cavalry operations, fell within our pre-set definition. There may not be a single reader of this book, in fact, who would have chosen the exact same table of contents. Frankly, we consider the subjective point of view to be one of the interesting aspects of this project, and, in any case, have no regrets that there are many more "great raids" yet to be explored in the future.

Some chapters that follow deal exclusively with a single raid, but others concern the raiders themselves—the personalities involved and the context in which they fought. Lawrence, for example, is not defined by Aqaba and Koos de la Rey is not defined by Tweebosch; these accounts, however, seek to impart not only their primary raids, but the type of fighters they were, and the impact they had on their respective conflicts. In planning the book, there was no thought given to whether a particular raid—or raider—was successful or not; our criteria was altogether different and had more to do with the attempt and the stakes involved than the reward.

In a book such as this, we are extremely grateful to a variety of people, aside from the authors, who lent their assistance on the subjects we were eager to pursue. To begin with, two of the pre-eminent experts in this field shared their thoughts and connections with us at an early and crucial point: Mr. Gary Linderer, founder and editor of *Behind the Lines* magazine, and Dr. John Partin, Special Operations Command Historian at McDill Air Force Base. These two gentlemen have our deepest appreciation for their enthusiasm and willingness to assist. Our sincerest thanks are also extended to Harry Cooper, President of Sharkhunters in Florida, Col. Michael E. Haas (USAF, ret.), Dr. Michael Palmer of the University of South Carolina and Col. Alexander McColl, special projects editor of *Soldier of Fortune* magazine. We also owe a debt of gratitude to Ms. Nancy Guier and Ms. Roxanne Merritt at the John F. Kennedy Special Warfare Museum, Fort Bragg, North Carolina, for photographs, and to Ms. Rita MacDonald for maps. Special thanks go to Col. James H. Kyle (USAF, ret.) for photographs of Desert One, and for having "the guts to try." Thanks also to Ms. Ann Crawford of the Public Records Office, London; Warrent Officer Smith of the Canadian Forces; Ms. Elsie M. Maddaus of the Schenectady County Historical Society; Ms. Lisa Wood of the University of Kentucky Library; and Maj. Andy Lucas (USA). The Kitzmiller family (John, Tamar, Willy and Hannah) of Norwich, Vermont provided morale and comfort to a frequent guest who spoke of nothing but raids. Wade G. Dudley not only contributed the opening chapter of this book but also brought to bear his immense knowledge of naval affairs as a consultant on other chapters. Jim Lawler offered invaluable expertise on European military history and Anne Dorothea Smith made available her wealth of knowledge on 19th century America and the Civil War.

Donn Teal, the rock against which all the waves that come through this company must break, stood steadfastly, red pencil in hand, copyediting the variety of chapters with his unfailing commitment to detail. Many thanks, too, to John Larsen and Erik Hildinger for their generous assistance. Finally, Lt. Commander Robert A. Southworth (USNR, ret.), should be thanked for telling the stories of Lawrence, St. Nazaire and Scapa Flow to an impressionable child during long ocean voyages in very small boats. "These should all be together in a book," thought the lad and, along with many others, now they are.

About the Authors

(in alphabetical sequence)

JAMES M. ALDRICH earned his A.B. from Harvard University and his Ph.D. from the University of Maine. His areas of expertise include U.S. maritime history and the Colonial era. For the past 12 years he has been an editor at the Penobscot Bay Press in Stonington, Maine. His previous publications include *Centennial, A History of Deer Isle As Told By 100 Years of Island Newspapers,* and *Fair Winds, Stormy Seas: A History of Maine Maritime Academy.* He and his wife live in Deer Isle, Maine.

JANIS CAKARS has spent his entire career in a variety of capacities in the field of military history publishing and is currently an editor at Sarpedon Publishers. A graduate of Hunter College in New York City, he currently lives with his wife, Melissa, in Brooklyn, New York.

GEORGE F. CHOLEWCZYNSKI, a former technical librarian, is an expert on airborne warfare and Polish military history. He is the author of *Poles Apart: The 1st Polish Airborne at the Battle of Arnhem* (Sarpedon, 1993). He has edited the memoirs of the late Colonel Witold Urbanowicz, who commanded the famed Kosciuszko Squardron and flew in China with the Flying Tigers, scheduled for publication in spring 1997. He and his wife Marie reside in New Orleans, Louisiana.

MICHAEL F. DILLEY is a former paratrooper, serving in the XVIII Airborne Corps and the 82nd Airborne Division. He retired from the U.S. Army in 1984 after duty as a counterintelligence agent, intelligence analyst, case officer and interrogator. He served in Vietnam for two years. He has a B.A. from Columbia College in Missouri, and is a staff writer for *Behind The Lines* magazine. He also reviews books with special operations themes for *Behind The Lines* and *Infantry*. He is the author of *GALAHAD: A History of the 5307th Composite Unit (Provisional)* (Merriam Press, 1996) and, with Lance Q. Zedric, *Elite Warriors: 300 Years of America's Best Fighting Troops* (Pathfinder, 1996) He lives with his wife, Sue, in Davidsonville, Maryland.

WADE G. DUDLEY, following two decades with Proctor & Gamble, is currently a student in Maritime History at East Carolina University in Greenville, North Carolina. His main area of interest is Napoleonic-era naval history, particularly the War of 1812. He lives with his wife, Susan, and 17-year-old son Glen in Greenville. Another son, Billy, 19, is an engineer with the First Cavalry Division in Bosnia.

STEVEN HARTOV is a veteran of the Israeli Defense Forces parachute corps and military intelligence, whose first book, *The Heat Of Ramadan,*was described by Publishers Weekly as "The finest sort of espionage thriller." His second, *The Nylon Hand Of God* (William Morrow, 1996), is set in New York and the Middle East. He owes his thanks for assistance with "Green Island" to noted author and expert on worldwide special operations, Samuel M. Katz.

FRITZ HEINZEN's academic background includes doctoral work in the Dept. of Government and International Studies at the University of Notre Dame, where he was also the Coordinator for Interdisciplinary Nuclear Studies; an M.A. in Security Studies and International Economics from the Johns Hopkins Univeristy School of Advanced International Studies in Washington D.C. and Bologna, Italy; and a B.A. in History, German and Government and International Studies from Notre Dame, with additional academic work at the Universität Innsbruck. Aside from his work with numerous academic and public policy organizations, he is the Director of KrADeG Book Services, which reviews books for magazines and journals in the fields of military history, national security and foreign policy. He also hosts several weekly discussion shows on history and religious publishing on America OnLine

RICHARD L. KIPER is a retired Special Forces officer who holds a Ph.D. in History from the University of Kansas. He received his M.A. in History from Rice University and his B.S. from the United States Military Academy. He has taught at West Point and the U.S. Army Command and General Staff College at Fort Leavenworth. Currently on the faculty of Kansas City, KS Community College and American Military University, Dr. Kiper has presented several conference papers and published numerous book reviews and articles.

ROBERT E. KROTT, a former U.S. Army officer and paratrooper, earned a B.A. in history from Saint Bonaventure University before attending graduate school at Harvard University. A veteran of the war in Bosnia, he served both as a commando officer in the Croatian Army (where he received the *Zahvalnica* medal for combat), and as an infantry officer in the Croat-Bosnian Defense Force. He has been awarded the parachute badges of eight foreign militaries. A frequent traveler to Africa, he is currently a columnist and foreign correspondent for *Behind the Lines* magazine, and senior foreign correspondent for *Soldier of Fortune* magazine.

TIMOTHY J. KUTTA is a former Marine Corps officer and freelance writer specializing in military affairs. His articles have been published in *Vietnam, World War II* and *Aviation History* magazines, as well as *Military Technical Journal* and the *Marine Corps Gazette*. He is a contributing editor to *Command* magazine and an editor for *Strategy & Tactics*, which features his regular column "FYI." His books include *Gun Trucks* (Squadron/Signal, 1996) and the forthcoming (1997) *DUKWs*. His current project is *The U-Boat War, 1939–45*, which is scheduled for publication in 1998.

GEORGE F. NAFZIGER (Captain USNR, ret.) served as a Director of the Napoleonic Society of America and is currently a director of the Napoleonic Alliance and a Fellow of Le Souvenir Napoléonienne Internationale. Among his many important works are *Napoleon's Invasion of Russia* (Presidio, 1988), *Poles and Saxons During the*

Napoleonic Wars (Emperors Press, 1991), *Imperial Bayonets* (Greenhill, 1996) and *Napoleon at Leipzig: The Battle of the Nations, 1813* (Emperors Press, 1996). He currently lives in Cincinnati, Ohio, with his wife and children.

BRENT NOSWORTHY began his career as a designer of historical wargames and was the founder of Operational Studies Group, a company based in Montreal. His first book, *The Anatomy of Victory* (Hippocrene, 1989) is considered one of the classics on the Marlborough period while his second, *With Musket, Cannon and Sword: Battle Tactics of Napoleon and His Enemies* (Sarpedon, 1996) has received critical acclaim. While lecturing widely, and contributing to a variety of publications, he is currently working on a book concerning the combat experience and tactics of the American Civil War. He lives with his wife, son and daughter in Brooklyn, NY.

STEVEN M. SMITH is the president and founder of Sarpedon, a publishing house (established 1991) that specializes in military history. He currently lives in Long Island, New York with his wife, Susan, and daughters Emily and Anne.

SAMUEL A. SOUTHWORTH is a writer and editor at Sarpedon Publishers. After earning his M.A. at the University of New Hampshire, he taught writing there for three years before pursuing a career in publishing. Trained as a Wilderness EMT at the North Carolina Outward Bound School and a graduate of a winter course at the National Outdoor Leadership School in Wyoming, he has planned and led mountain and river trips for fifteen years. He currently lives in New York City and has written poetry, short stories, feature articles, textbooks and a novel.

STEPHEN TANNER is a full-time contributing and acquisitions editor of non-fiction history publications. His recent projects include a general history of aircraft and an account, originally in Dutch, of the career of Toussaint Louverture (optioned by Hollywood in February 1997). His military history projects have ranged from the Roman era through the Civil War to World War II. Based in New York City, he is married and has two daughters, ages ten and fourteen.

CHARLES WHITING is one of the world's bestselling authors of military history, whose many—and often controversial—works on World War II include *Bloody Aachen, Death of a Division, The Search for Martin Bormann, The Last Assault: The Battle of the Bulge Reassessed* and *Death on a Distant Frontier.* His books have been translated into a dozen languages and many continue to be republished in paperback. He currently lives in York, England.

W. RAYMOND WOOD, Ph.D., is a Professor of Anthropology at the University of Missouri, a past editor of the journal of the Society for American Archaeology, and has written numerous books and articles on the culture and history of the American West. He is also the author of *Or Go Down In Flame: A Navigator's Death Over Schweinfurt* (Sarpedon, 1993), the result of his years of research into the fate of his brother, Elbert Stanley Wood, who was killed in action on October 14, 1943 over Germany.

Introduction

BY SAMUEL A. SOUTHWORTH

Military science is a multi-faceted discipline that simultaneously examines the past and attempts to glimpse the future. It is studied by historians, professional military planners and a multitude of people whose curiosity about the very limits of human behavior—including creative and physical skill, ferocity and courage—never ebbs. Each in their considerations must take into account the various commanders, terrain, morale, logistics, weather, time, technology and, not least, the element of chance. There are strategic concerns that must be weighed against tactical methods of achieving goals: long campaigns that drag on through months of fighting, interspersed with tedium; set-piece battles that are almost formal in their arrangement of opposing forces; huge invasions that rumble onto hostile shores; as well as the sweep of mighty armies, indicated with red and blue arrows on maps that cover entire continents. And then there are raids.

The raid as a military operation is the second most primitive tactic in the history of warfare, after the ambush. Yet, as this book demonstrates in its progression through the last four centuries, it has also become the most sophisticated type of operation in the modern era, requiring the most brilliant commanders and the most elite soldiers. While tribal or primitive societies may have practiced raiding in the past as a substitute for the ability to maintain conventional armies—and, as a rule, either against equally primitive opponents or in efforts at strategic defense—the history of raiding since the inventions of gunpowder and long-range weapons describes sticking one's head into the tiger's mouth *instead* of conventional battle, pitting small forces in pursuit of limited goals, where they face the risk of annihilation.

In between the period when most of the world, outside of a few regional empires, was still tribal or nomadic in nature, and the subsequent development of nation-states with fixed boundaries, there were centuries of transition when kingdoms and fiefdoms shifted borders at the stroke of a sword, and raiding was a common form of warfare. The Vikings, for example, had raiding as a cornerstone of their culture, as, prior to Genghis Khan, did the steppe warriors of central Asia. The Knights Templar protected their sovereignty in the Mediterranean, after the Crusades, by constant thrusts against the surrounding world of Islam, until finally forced to stand against those armies at Malta.

This book begins with the events of 1587, after Europe had politically evolved into a system of states, each with standing armies, some much stronger than others. By this time raids, as a means of waging war, were not the standard methods of tribes or peoples, but the option of the most inventive and daring members of a country's military forces. As armies became better trained and more powerful, and technology began producing ever more deadly weapons—capable of greater firepower at ever-increasing distances—the notion of raiding became an increasingly dangerous proposition.

In the following pages various patterns emerge, as the chapters wind their way from the Elizabethan era through the presidency of Jimmy Carter. In a technological sense, the view is clear: windblown sailing ships give way to motorized landing craft and SCUBA assault vehicles; horses (or camels) become troop-carrying aircraft. Weapons progress from smoothbore cannon to air-to-air rockets. Another pattern that emerges is of a geopolitical nature: alliances shift, and nations that had been enemies for a generation can, in a subsequent chapter here, suddenly be recounted as friends. Perhaps Britain's is the most consistent presence in these chapters, standing as it has at the fulcrum of great power conflict throughout the modern era.

But evolution, whether in technology or politics, is not at all the entire story of the modern history of raids. When considering the personalities, one can conjure an entire kaleidoscope of scenarios: if Francis Drake had been an officer of Tennessee cavalry during the Civil War; if Yoni Netanyahu had been the captain of an American frigate in 1778; if the Cossack Czechninev were a British commando;

or if Custer had been an Israeli in the 1970s. In one sense, the history of raids in the modern era is amazingly consistent. The kind of commanders who took enormous risks on behalf of their countries, in desperate, "winner take all" ventures, has not changed an iota in the past 400 years. Audacity will never fail to astonish, and a bold leader who has earned respect will always be followed, even if the proposed action is not only hazardous but seemingly impossible. Raiders, be they "sea dogs" or Green Berets, rely on the simple formula of outstanding leadership to execute startling plans.

As for the concept of "great raids" in history, the reader will recognize that this book includes a number of unmitigated disasters, although none without displays of bald courage. Raids, in fact, often fail. In other cases (Deerfield, Morgan, de la Rey, Chindit), this book describes cases where raiding as a military technique is employed to wear down the enemy with repeated incursions into his territory. The men may succeed, but the "cause"—to succeed, if at all—must wait for greater, conventional, strength to be brought to bear.

Raids, in themselves, do not win major wars. However, as wars between great powers become less common, the ability of powerful armies to stage raids (the Congo, Desert One) becomes even more central to achieving immediate national goals. There is a degree to which, as we approach the year 2000, the lessons we have learned from the great battles in history—Waterloo, Gettysburg, Normandy—have already become anachronistic, in view of the fragmented world in which we now live. One must consider the possibility that no one nation's cause can any longer be won by the massing of huge amounts of men and materiel; the decisions in the future will instead be won by the most daring commanders, albeit with the best weapons available. These leaders, and the brave soldiers who choose to follow, may well decide the military history of the future.

This book, in describing some of the raids that have taken place throughout the modern era, presents no particular theme on the subjects of weapons technology or geopolitical alliances. It does intend, however, to reveal something of the nature of the soldiers themselves who have most courageously served their countries, across centuries of Western history. There are certain "constants" in the world, and these, having mostly to do with character, are illuminated in the following pages.

1. Drake at Cádiz
2. The French and Indians at Deerfield
3. John Paul Jones off Britain
4. The Cossacks at Hamburg
5. Mosby at Fairfax Court House
6. Morgan Across the Ohio
7. Custer at the Washita
8. Koos de la Rey in the Transvaal
9. Lawrence at Aqaba

10. Gunther Prien at Scapa Flow
11. The British at St. Nazaire
12. The Canadians at Dieppe
13. First Chindit
14. Second Schweinfurt
15. Skorzeny at Budapest
16. Belgian Paras in the Congo
17. The Israelis at Green Island
18. The Americans at Son Tay
19. Delta Force at Desert One

GREAT RAIDS IN HISTORY

"When it comes to the pinch,
human beings are heroic."
—George Orwell

1

Drake at Cádiz

BY WADE G. DUDLEY

If a gull could have hovered in that wind-torn sky, it would have observed twenty-three small ships spread across several hundred square miles of tossing sea. The waves towered over them, squalls of rain blinded their helmsmen, canvas ripped and was replaced. A tiny pinnace lost way, broached, and was swallowed by a mountainous wave, leaving no marker for its damp grave. Each ship fought its own battle against the uncaring sea. Perhaps the gull would have noticed one ship forging far ahead of the others, as if driven by demons. Possibly, it would have landed there, on the poop, resting until startled by the cry of a commanding voice.

"Reef the t'gallant! Take in sail!" With that, the short, rather stout seaman dropped his line and staggered toward the wheel. There was a wild sea running; it had been that way for four days. The quarter-master needed all the help available to keep the *Elizabeth Bonaventure* from broaching, and drowning the lot of them. As he grabbed the spokes of the wheel, his beard dripping from the blown spume and pouring rain, the squat seaman laughed aloud.

The other English sailors looked at the man. As always, their dedication to him, their respect for him, only intensified; for he was more than a simple seaman. Though he pulled a line, reeved new tackle, or steered the ship alongside each of them when needed, he was also, by the Queen's grace, their admiral. He was Sir Francis Drake, a gentleman willing to stoop to common tasks and a common man able to stretch to heroic proportions. One can almost hear the echo of his crew's words, "See how he laughs at the gale! He fears only God!"

If Sir Francis laughed, it was because his greatest fear was behind him. The storm spelled the end of any recall attempt by the Queen.

True, his fleet was scattered, but they would make the rendezvous or not, as God willed. Then the campaign would continue. If Elizabeth vacillated, as she often did, well, not even the Queen could find his fleet now. So God had willed it! Though Drake would never have compared his skill at statecraft to that of Elizabeth, he trusted his own ability to see the trees despite the forest. And those trees had led him to that gale in the Atlantic.

Throughout the reign of Elizabeth, the relationship between England and Spain had been menacingly tense. At its heart was religion. England was a bulwark of the Protestant movement while Spain was the champion of Catholicism. It is difficult for those of us living in an enlightened America to truly comprehend the depth of that conflict, where men who worshipped the same God, but in a different manner, daily burned, tortured and killed each other simply because of their beliefs. Atrocity bred hatred; hatred guaranteed that one faction must try to convert, or destroy, the other. By 1586, Protestants in the Netherlands were under military pressure from the armies of King Philip II of Spain. The Dutch were supported by English military forces and English gold. Philip recognized that only an invasion of England, and its restoration to the Catholic fold, would allow the triumph of the Church (and Spanish hegemony) on the continent.

Philip ordered the formation of a vast Armada. New ships were to be built and others gathered from client states or impressed from merchants. Weapons were to be found for an army. New cannon were to be cast and rushed to the ships. Provisions were to be gathered from far and wide—half a million pounds of cheese from the Baltic, rice from Milan—and thousands of wooden casks were to be built to hold these provisions and water for the fleet.

By the spring of 1587, it appeared that the Armada could sail by late summer, delivering its army to English soil before the unpredictable fall weather made a seaborne invasion too hazardous. There were also other reasons for the Armada to sail on time. Provisions spoiled rapidly with only salt to preserve them, and would not keep for more than half a year. Failure to sail would also mean the continuing concentration of large numbers of men in the close quarters of Spanish ports. Aside from the cost to the royal treasury, disease and epidemics were likely to break out and the strength of the army would erode.

Drake, as a seafarer, understood the timetable that the Armada faced. Drake, as an ardent Protestant, also understood the danger to England. He had lived through the reign of "Bloody Mary" before she was toppled and Elizabeth made queen; he had known men martyred for their resistance to Catholicism. Drake, as a seafarer, understood the enemy's timing; as a Protestant, he had motive. But it was Drake, the veteran raider, who time and again in the winter of 1587 approached Queen Elizabeth for permission to mount a pre-emptive strike at the Spanish mainland.

Elizabeth hoped for peace. She had grasped at every straw, short of marrying Philip, to procure it over the years. The Armada seemed to have been less than real to her, some vague nightmare looming on the horizon. England's involvement in the defense of the Netherlands, however, provided a more pressing motivation. Her state coffers were drained almost to the bottom. After weeks of pleading for a mission, perhaps Drake finally found the magic word to say to Elizabeth: treasure. Drake had seldom gone into action against the Spanish without returning laden with gold and goods, much of which in this case could easily help refill the English treasury. In any event, Drake won Elizabeth over and he received his commission on March 15, 1587.

His orders called for a four-prong mission: to disrupt Spanish shipping between the Mediterranean and Spanish-controlled Lisbon; distress enemy ports if possible; seize ships from the West and East Indies; and harass the Armada should it already be at sea. Such orders were a raider's dream! They offered tremendous flexibility and no limits to initiative or action. On the other hand, Drake realized that his orders represented a veritable declaration of war against Spain. He worried constantly about the Queen rescinding them.

With that concern in mind, Drake rushed to gather a fleet. As with most efforts of those days, the expedition was organized as a joint-stock company, shares of plunder going to each investor based on ships and men committed. Drake, already famous for his adventures in the New World, had no problem finding investors. In the end, the Queen provided six ships, totaling 2,100 tons and 1,020 men; London merchants contributed twelve ships, totaling 2,100 tons and 894 men; Drake himself provided four ships of 600 total tons and 619 men; finally, Lord Admiral Howard provided one 175-ton ship and 415 men. Such was Drake's rush to be away, that all the ships were gathered and provisioned by April 1, 1587. The next day twenty-three ships and nearly three thousand Englishmen set sail for Spain—for God, and for glory.

From the cabin of his flagship, *Elizabeth Bonaventure*, Drake wrote to minister Francis Walshingham as the fleet cleared the harbor, "The wind commands me away. Our ship's under sail. God grant we may live in His fear as the enemy may have cause to say that God doth fight for Her Majesty as well abroad as at home . . . Let me beseech your honour to pray unto God for us that he will direct us the right way . . . Haste!"

"Haste" had been a well-chosen tactic for Drake, his fears that Elizabeth would undergo a change of heart well founded. Even as the ships sailed from Plymouth, the Queen's messenger galloped from London to recall the fleet. He arrived too late. The pinnace dispatched in pursuit of the fleet likewise failed to find Drake, perhaps because of the storm, more likely because of the five-thousand-pound prize that they stumbled across and captured. Regardless, it was a stroke of purest fortune for England that Drake continued undisturbed.

And so Drake could laugh in the face of an Atlantic gale. His contemporaries wrote that he was always quick to laugh. Perhaps the trials of his youth had taught him to laugh at adversity. His family had been reduced to poverty during the Catholic reign of Mary. His father had given Francis the rudiments of education, had taught him to be a God-fearing Protestant, and had apprenticed him to the master of a small coastal trader at the age of thirteen. It was here that the young man learned to love the sea. It proved to be the greatest of highways to adventure, wealth, and fame. The sea taught Francis Drake that the tide never waits; once missed, forever missed. Opportunities must be seized immediately.

By 1587, Drake was in his late forties. He was described as "low of stature, of strong limb, round-headed, brown hair, full-bearded, his eyes round, large and clear, well-favored of face, and of a cheerful countenance." Short he may have been, but by that year he had proven himself a seaman-raider of greater stature than any other. Drake had tweaked the nose of the Spaniards throughout the Caribbean, capturing ships and looting towns. He had captured a Spanish treasure galleon and ballasted his ship with silver ingots. He had circumnavigated the world, the first Englishman to do so. He had been knighted for his deeds and walked close to his queen. The name of Sir Francis Drake was known throughout Europe, and feared throughout Spain.

On the fifth day, the storm ended. The fleet straggled to the rendezvous at the Rock (Cabo Roca, near Lisbon, Portugal). The gale had been weathered surprisingly well, though the queen's ship *Dreadnought* had developed a leak and a pinnace had disappeared.

It is not certain that Drake, the eternal opportunist, had decided how to proceed with his mission at that time. Perhaps it was a case, as he so devoutly believed, of "God will provide." Regardless, several Portuguese prizes were taken, at least one of which was crewed by a few Englishmen and these remained with the fleet. More important, Drake bespoke a Flemish merchantman recently sailed from the port of Cádiz in southern Spain. Its master told him that the harbor was crowded with shipping: ships of the Armada, supply vessels for the Spanish endeavor, and merchantmen. Thus, in the mind of Drake, was birthed the raid on Cádiz .

Drake quickly called the traditional council-of-war aboard the *Elizabeth Bonaventure*. His second-in-command, Vice-Admiral William Borough, was a good navigator of great fame, but he was not a warrior. He counseled caution, little knowing that it was said of Drake, "Though a willing hearer of other men's opinions, he was commonly a follower of his own." And his own opinion was that Cádiz was ripe for a raid. As Borough prepared to leave for his ship, he asked Drake for orders. The reply was, "Follow your flagship," and if Drake had only considered, he might have added, "and do what must be done!" With captains again aboard their ships, the fleet spread sail for Cádiz .

On April 19, the fleet arrived off the Spanish port. By naval standards of the day, it was time to heave to for a final council-of-war, to allow the slow-sailing stragglers to close with the fleet, and to reconnoiter the unknown harbor. Borough was horrified to see Drake not even slowing as he sailed directly for the narrow harbor entrance. The two forts guarding the entrance were clearly visible, and soon a mass of shipping in the outer harbor could be distinguished, protected by two galleys.

These fast and agile oared ships with their heavy bow gun and several man-killers, packed with cutthroat Spanish soldiers, were the terror of the Mediterranean. They did not need the wind, and could lie off the bow or stern of a becalmed sailing vessel and pound it into bloody submission, or even ram with their steel-shod bows. Borough knew that galleys would have the advantage inside the port, as the surrounding land smothered the wind.

Drake had fought the Spanish all of his adult life. His decision to quickly enter the harbor was based on his belief that Spaniards were never prepared for battle; if, by some miracle they did react quickly, their powder was unevenly corned and always lacked accuracy. As for the galleys, Drake's fleet had wind, and was made of timbers stout enough to defeat Atlantic storms. His well-served, long-range cannon would crush them at a distance. What was needed was surprise! Sail into the harbor as if he owned it, and the Spanish would be so easily demoralized as to offer little resistance.

So the fleet stood in, no flags flying, no drums beating. By four in the afternoon, Sir Francis Drake was within the outer harbor of Cádiz . The galleys, reinforced to six, approached to investigate. Up flags! The lion of England flew proudly as the four lead ships offered their broadsides to the approaching galleys. Their fire so damaged one that it was beached, sent another limping for the inner harbor, out of action for the near future, and caused the remainder to flee for their fragile lives.

In the city there was horror. "The Dragon Drake!" people cried, that Protestant spawn of Hell who ate Spanish plunder and burned their ships with his demon breath. Now he would treat Cádiz as he had the towns of the Spanish Main! Panic ensued as citizens rushed to the city's citadel for safety. The terrified commander shut his gates in their faces. Twenty-six women and children were crushed by the stampeding mob. It was only with great difficulty that the local general managed to restore order; it was only as dusk fell that he managed to place garrisons at the most likely landing sites along the waterfront, commanding his remaining eight galleys to concentrate there. His messages requesting reinforcements would be answered by the Duke of Medina Sidonia, future commander of the Armada, but help was hours away, and Cádiz ripe for the torch that very minute. It was fortunate for the harassed officer that Drake's interest was focused elsewhere.

As Drake's fleet gathered around him in the outer harbor, the chaos in the city was reflected by the panic of the shipping crowding the roadstead. Of the sixty ships peacefully at anchor minutes before, some twenty fled upriver to the port of Santa María. A few were able to creep into the inner harbor. Most simply milled in panic, masters cursing and hulls crashing together.

Then from out the flock of terrified sheep emerged a wolf. A thousand-ton behemoth, almost twice the size of Drake's largest ship, decided to make a fight of it. It would be the only ship to do so. Drake would have dearly loved to add such a monster to his fleet, but when its forty guns began to speak, his options became limited. The deadly accurate English guns answered back. Interestingly enough, this one ship to fight for the harbor of Cádiz was not even Spanish, but Venetian; and why a captain of Venice would have chosen to die in defense of a foreign port we will never know. By dusk, ship, crew, and captain were at the bottom of the harbor.

Dusk gave way to a cold, dark April night. Drake was master of the outer harbor, and the true work of the English now began. The ships' boats made their way to vessel after Spanish vessel. Soon the night grew bright as these began to burn. Perhaps some Spaniards begged that their means of living be spared; probably there was the glint of firelight on knives, and the song of steel briefly sang; it did not matter. Ship after ship was shorn of its portable wealth, four vessels loaded with provisions were taken prize. The remainder were put to the torch. Incomplete warships for the Armada, storeships of provisions and naval stores, and unlucky merchantmen all warmed that cold April night. A Spanish soldier, one among the six thousand troops brought to Cádiz by Medina Sidonia, captured the essence of the scene: "With the pitch and tar they had, smoke and flames rose up, so that it seemed like a huge volcano or something out of Hell . . . A sad and dreadful sight."

As the burning progressed, Borough made his way to Drake's flagship, with the intention of drumming some sense into him. Their actual conversation was not recorded. Undoubtedly, Borough counseled that success had been great, and they should withdraw with the morning light before the Spanish could muster real resistance. Apparently, he returned to his ship feeling confident that Drake had listened. Imagine his disbelief as the light of dawn revealed one of the smaller English ships, the *Merchant Royal*, at the very entrance to the inner harbor while a flotilla of ships' boats were rowing madly into it!

Amid the refugee shipping clustered within the inner harbor, Drake had spotted the twelve-hundred-ton galleon that was being fitted out as the eventual flagship of the Armada. It had to be taken (recall that his investors required a profit) or destroyed. Ordering the *Merchant Royal* to provide covering fire against roving galleys, he personally led a cutting-out expedition into the inner harbor. Finding the ship lacking cannon and sails, Drake reluctantly ordered it burned. The scene of the previous night was repeated, as the galleon and all merchant ships within easy reach became ashes wafting over Cádiz .

Borough, meanwhile, had taken to his boat in search of his admiral. This was madness, the fleet had done enough! It must sail now! He missed Drake at the flagship, and again at the *Merchant Royal*, which he ordered to retire somewhat, for the sake of safety. By the time Borough returned to the *Elizabeth Bonaventure*, Drake was back. Their confrontation, past the stage of giving and receiving counsel, found Drake speaking disparagingly of Borough as "in trembling sort." For Borough's part, he felt himself "in haste" for the safety of the fleet rather than fearful for himself.

Returning to his ship, the *Golden Lion*, Borough found that it had been hulled below the waterline by a shot from a new battery. This, apparently the only successful shot of the entire raid by a shore gun, broke a gunner's leg. Borough withdrew some five miles closer to the harbor mouth, ordering several ships to support him as he did so. Drake, though puzzled by the move at the time, saw that the new location would cover the fleet's withdrawal from the pesky galleys. He ordered Borough reinforced to a strength of six ships.

The galleys—a total of ten would be involved in the latter stages of the battle for the harbor—proved to be at best a nuisance to the English fleet, despite the continued apprehensions of Borough. Their vaunted agility availed them little, since they could not approach the massed guns of the English and remain afloat. Their one success came when the Portuguese merchantman captured enroute to Cádiz was isolated by several galleys. Even then, the vessel was not taken until all of its five-man crew lay dead or incapacitated. Surrender was not counted an option when the most likely fate of prisoners would be to live their remaining days as slaves in those same Spanish galleys.

The work in the inner harbor was complete by noon. Drake

ordered his fleet to form in line and head for the harbor's mouth as ashes fell like snow. The wind picked that time to die. One can almost hear the I-told-you-so's of William Borough ringing across the vault of history. It did not matter. Drake formed his ships for defense from all quarters, and when, with bulldog-like persistence, the galleys attacked, and attacked again, they were beaten off. With the fall of night, the Spaniards sent fireships against the fleet. Drake sent his pinnaces and smallcraft to fend them away, laughing at how the Spanish dogs continued the English work by burning their few remaining ships. At two in the morning on April 20, 1587, the wind returned. Drums beating, horns blaring and flags flying, the fleet of Sir Francis Drake sailed from the ash-covered bay of Cádiz .

By Drake's own reckoning, he had sunk two galleys and the Venetian galleon. His forces had burned thirty-one vessels of two hundred to one thousand tons, as well as a twelve-hundred-ton galleon. Four vessels had been taken as prizes, aside from enough provisions to replace those expended thus far in his voyage. Other portable plunder taken was never cataloged, nor were the vessels used as fireships by the Spanish ever counted. The English loss had been five dead or captured with the Portuguese vessel, and one broken leg. Spanish sources reported fewer ships lost, only twenty-four or twenty-five, at a cost of 172,000 ducats (somewhat over ten million dollars in today's currency) to Philip. Like the actual amount of booty seized by English sailors that they were required to report to the fleet's investors, those numbers were most likely conservative.

In later years, a laurel-laden Drake would think back on that flaming night in Cádiz , laughing as he recalled his "singeing of the King of Spain's beard." On the night of April 20, however, his thoughts were elsewhere. The mission was not complete. A blow had been struck against the Catholic foe, but the Armada was far from crippled. Cádiz was behind him, but the raid was not complete.

His immediate problem was to create a safe haven. His men were beginning to sicken from scurvy, because all their fresh vegetables were long since gone, and probably from the fevers that came aboard with body lice whenever ships touched shore. The sick needed a brief respite on land. Fresh water was also a concern.

On May 3, 1587, Drake ordered a subordinate to seize Lagos, at the base of Portugal's Cape St. Vincent. He failed. Two days later

Drake personally led an assault on the key fort defending Sagres, near the tip of the cape. Under heavy fire, he helped pile combustibles at its wooden gate and began to burn his way in. The garrison, their officer wounded in two places, surrendered. The three remaining fortifications at Sagres, demoralized by the Dragon Drake, capitulated immediately. Drake ordered all guns thrown into the sea and razed the fortifications.

After resting his men, Drake terrorized the coast of Portugal. His new base was ideally located to intercept traffic between the Mediterranean and northern Spanish holdings. Ship after ship was seized, stripped, and burned. In Drake's own words, ". . . it hath pleased God that we have taken forts, ships, barks, carvels, and divers other vessels more than a hundred, most laden, some with oars for galleys, plank and timber for ships and pinnaces, hoops and pipe-staves for cask, with many other provisions for this great army." He estimated that seasoned wood sufficient for casks of "25 or 30 thousand ton" had been destroyed.

Consider the impact. If the Armada sailed, it would do so with leaky barrels made of green wood. Water and provisions would be contaminated. Crews would quickly sicken and be incapacitated. Perhaps the single greatest blow of Drake's extended raid was the destruction of those pipe-staves. When the Armada finally sailed, in 1588, it did so with barrels made of green wood.

Holding Cape St. Vincent was not enough for Drake. For two days he blockaded Lisbon, assembly point of key elements of the Armada and its admiral, Santa Cruz. He captured merchantmen, or ran them ashore. He made target practice of the handful of galleys that ventured to oppose him. Day and night his drums beat their war cry, his trumpets blared their challenge. Drake even sent taunting messages ashore—all to no avail. Santa Cruz would not fight, because he could not. Spars, sails, barrel staves, and munitions had already felt the breath of the Dragon; his ships lacked the ability to sail.

Surrendering only to fate, Drake returned to his base at the cape. Enroute he stumbled on the local tuna fleet of fifty vessels. Drake took their catch, destroyed their nets, and sank most of the boats, excepting those needed to set the fishermen ashore. If the villagers must starve that year so that the soldiers of the Armada might suffer from a shortage of food, so be it. War was harsh on everyone. Let it be harshest on the Catholics!

If holding Cape St. Vincent was not enough for Drake, then it had proven too much for Vice-Admiral Borough. In a critical letter to the admiral dated the day before the capture of Sagres, Borough challenged Drake's leadership of the fleet and his unwillingness to take counsel from his second-in-command. It was time to sail for England. Drake's fury knew no bounds. It was not a time for caution! It was not a time for divided leadership! And no right time ever existed to criticize the headstrong Drake to his face. Admiral Drake stripped Borough of his command and ordered him confined in his cabin until the return to England, where he could be tried for cowardice.

By the third week of May, shipping had disappeared from the coast of Portugal. Masters refused to challenge the odds in favor of their destruction at the hands of the English. Better to rot in harbor. But Drake faced his own problems. Scurvy and fever had scourged his crews. Ships were beginning to show the wear and tear of the Atlantic breakers. Retaining only his ten best ships and most healthy crewmen, he dispatched the remainder to England on May 22, 1587. The targets of the raid had disappeared, but the raid was not quite over.

Drake and his ten ships disappeared into the Atlantic. King Philip's brow, above his much singed beard, furrowed with consternation. Bad enough to know exactly where Drake was burning his shipping, but now where could he be? There was only one answer. The Treasure Fleet! Millions of pounds in silver, all desperately needed by the near-bankrupt Spanish treasury, sailing on a well-known course from the New World to Cádiz . Philip ordered every warship capable of bearing sail into the Atlantic. They would not find Drake, and, in fact, the Treasure Fleet would evade his depredations. But by the time the Spanish ships returned home, with worn vessels and sick crews, the opportunity for the Armada to sail in 1587 no longer existed. Drake had bought a year for Elizabeth and England simply by disappearing.

Where was Drake? Looking for treasure, of course, though not for the huge fleet then en route from the New World. His remaining vessels were not strong enough to overcome more than a single galleon. By the time he reached São Miguel in the Azores, the fleet was down to nine ships. The *Golden Lion*, former flagship of Borough, had mutinied. Its captain was put overboard in a small boat to inform Drake that crew and ship were sailing for England. Borough's complicity in the plot was never proven, though Drake tried him in absentia and condemned him to death.

Drake lacked the time to worry long about Borough. His raid had been a brilliant success thus far, but it still lacked a suitable return on his investors' money. On June 9, that changed. A large sail had been sighted the evening before and, as the morning mists cleared, a large Spanish galleon appeared. It was the *San Felipe*, mounting over forty guns to protect her cargo, a cargo hauled from the faraway East Indies.

With three of his ships, Drake stood directly for the towering galleon. No flags were flown, and the decks appeared as disordered as any Spanish warship. By the time the captain of the *San Felipe* recognized the ships as English, it was too late. Like terriers they ranged along her flanks, decks so low that the galleon's guns could not depress to reach them. Ball after ball plowed into the behemoth, carefully angled up so as not to sink it. The contest raged hotly until the captain of the galleon fell, severely wounded. As often proved to be the case, the Spanish sailors quickly surrendered once bereft of leadership.

Imagine the joy of the English sailors to find their prize loaded with the silks and spices of the Orient; and crammed wherever space allowed was the cargo of a second galleon that had developed a leak at the Spanish ship's last port! For the crew, these were riches beyond their wildest dreams. For Sir Francis Drake, there was the satisfaction that his investors would receive shares beyond their imagining (112,000 pounds total value, or about five million of today's dollars) and that his mission was at last completed.

On June 26, 1587, Drake's fleet dropped anchor at Plymouth, from whence it had sailed only eighty-five days earlier on one of the greatest raids ever recorded. Drake and his crews had boldly entered a defended enemy port and destroyed its shipping. They had razed enemy fortifications, establishing a temporary base on enemy soil. They had sunk, burned or taken over 150 enemy vessels, many loaded with war materials. Drake had even dared to taunt the commander of the Armada within his own home port. When the British fleet disappeared, every Spanish warship had scurried to find it, yet Drake still found a rich galleon, captured it and sailed through the Spanish searchers to return home. Of such threads are legends woven.

The impact of Drake's raid far exceeded the damage to the gathering Armada and its delayed sailing for a year. The staunchest of Catholics, the Pope, wrote in the year that the Armada finally sailed, 1588, "Just look at Drake! Who is he? What forces has he? And yet he burned twenty-five of the king's ships at Gibraltar, and as many again at Lisbon. He has robbed the flotilla and sacked San Domingo. His reputation is so great that his countrymen flock to him to share his booty. We are sorry to say it, but we have a poor opinion of this Spanish Armada, and fear some disaster." The final and most important gift of Drake's raid was the splintering of Catholic solidarity.

A raid, by definition, is the application of limited force to achieve specific objectives behind enemy lines or within delineated enemy territories. Not so clearly defined is the fact that raids are "do or die" missions. Failure finds the losers isolated, with the choice of death or surrender. Unfortunately, success often leaves raiders with the same option.

This makes Drake's raid on Cádiz and the aftermath a classic example of success. Not only did Drake accomplish his main missions—to disrupt shipping, distress the enemy in his ports and capture ships from the West and East Indies—he lost but five men to enemy fire. The remainder of his men (those surviving disease) returned home, and did so with the richest prize ever captured. Why?

The answer resides in the character of Sir Francis Drake. If the reader disagrees, then picture Borough in charge of the raid. There would not have been a Cádiz , much less an aftermath. (For the sake of closure, Borough was found innocent of charges by Queen Elizabeth. During the invasion of the Armada, he captained a barge on the Thames; soon after he drifted into obscurity.) Borough may well have been a better mariner than Drake, but he lacked the qualities that made Drake a successful raider.

Francis Drake not only took risks, he had the ability to inspire men to match his daring against the enemy. He was Protestant to the heart, and despised Catholics, particularly the Spaniards who supported Bloody Mary in her ascent to the throne of England. Drake had an eye to the main chance, an uncanny sense of judgment, born of harsh experience and intuitive skill, that so often led him to the right place at the right time that the Spaniards thought the devil whispered in his ear, telling him where their ships cruised.

Drake led men; he did not manage them any more than he managed his battles. He worked beside them, fought beside them, shared their humble fare; yet when he commanded, they obeyed. The skill of true leadership, being part of and still separate from those led, is a difficult art to master. It remains a distinguishing mark of the greatest of raiders. That Drake and England benefited from his leadership of the fleet is reflected in a sentence from a letter written by a subordinate shortly after Cádiz : "We all remain in great love with our general and in unity throughout the fleet."

Drake was never a strong planner. His raids would begin haphazardly, sometimes even running low on provisions shortly after leaving port. On the other hand, he had a strong sense of mission. One can almost envision him checking things off on a mental tally as the raid progressed: Harass enemy port—Cádiz; disrupt enemy traffic—Sagres; harass Armada—Lisbon, but Santa Cruz will not play; treasure ships—*San Felipe*; mission accomplished—go home. Drake could have stopped at any point advised by Borough with no failure attached, but his sense of mission was too strong to see any opportunity missed. That sense of mission—the ability to get the job done despite bad advice, despite mishap and despite distractions—is another hallmark of the successful raider.

These qualities made Sir Francis Drake a legend in his own time. When he departed the earthly stage he stepped beyond history and into English myth. His raids went with him, inspiration to the generations of raiders that followed.

The poet Greepe captured the feelings of that era towards Sir Francis Drake. He penned these lines in 1587 to celebrate the hero of the year, little knowing that they celebrated the greatest raider of the ages. They are a fitting ending to this story.

> Ulysses with his navy great
> In ten years' space great valour won;
> Yet all his time did no such feat
> As Drake within one year hath done.
> Both Turk and Pope and all our foes
> Do dread this Drake wher'er he goes.

2

The French and Indians at Deerfield

BY STEPHEN TANNER

The northwest frontier town of Pocumtuck was hard pressed by the forces of King Philip in the summer of 1675. Northfield, the only community situated farther from the center of the colony, had already been attacked, abandoned by its garrison and then burned, its traces disappearing into the soil from which it had just recently arisen. Pocumtuck's population, of not more than 200 transplanted English, was then reinforced by soldiers, because Pocumtuck had now become the "front line" of the war.

King Philip, a Wampanoag, was leading a coalition of tribes in an offensive against the English which, for its astute strategy and effective tactics, was unlike anything Europeans in the New World had yet seen. On September 1, a soldier was killed while outside the Pocumtuck stockade, and the Indian attackers herded the remaining garrison inside while they burned seventeen outlying buildings. On the 12th, twenty-two men attempting to reach a neighboring town were attacked, another man killed, and all were forced to seek refuge at their destination. The remaining men in Pocumtuck hunkered down helplessly as the Indians carried off harvested food and livestock, the remaining garrison too weak to interfere.

By September 18, the English had reasserted control of Pocumtuck and its environs; however, as a prudent measure, they decided to remove the thousands of bushels of corn that remained in the fields from the recent harvest. There were some fifty soldiers and armed militia present, aided by fifteen Pocumtuck men who drove the wagons in the column that headed south, toward the nearest town,

Hadley. While crossing a small stream, the column no doubt slowed, and perhaps became congested, and that is when over five hundred Indians suddenly attacked. In the ensuing battle of near-total annihilation, known as Bloody Brook, sixty Englishmen were killed, including all but one of the Pocumtuck teamsters. The next day, Indians danced outside the beleaguered town, waving clothes taken from the English bodies. Twenty-seven men remained within the stockade, helpless once more to intervene. Within days, Pocumtuck was abandoned, and then burned by its conquerors, just as Northfield had been. When settlers came again, to rebuild a community from the ashes of its doomed predecessor, they also changed its name to what it has been known as ever since: Deerfield, Massachusetts.

King Philip burned or destroyed twelve towns in New England, forcing five to be abandoned, and cowed many others in 1675, driving back the English encroachment onto traditional Indian land some 25 miles. Although Philip's father was a peaceful man, who may well have attended the original Thanksgiving dinner, Philip (whose birthname was Metacomet) had come to realize that, through their generosity and accommodation toward the English, the Indians had created a monster. He secretly organized a coalition of tribes, and then, when a spark set off the conflagration, launched his men in a series of attacks designed to seize territory and amputate the growing network of English settlements.

Even while New Englanders reeled from the devastating attacks on their outermost towns, within the colony they had no clear idea of which local Indians were for them or against them. The Narragansett tribe seemed particularly ambiguous, and in all probability, they thought, secret allies of King Philip. In December, an army of a thousand men—from the Massachusetts, Connecticut and Plymouth colonies—fell on a fortified Narragansett village in what was called the Great Swamp. After a hard battle, over 200 English were casualties but 600 Indians had been killed, about half of them warriors. In addition, the English burned the Indian town and all its food that had been stockpiled for the winter, rendering the tribe destitute. The English action also ensured that whatever Narragansetts remained would henceforth certainly be their sworn enemies.

By the spring of 1676 the English were better prepared for renewed attacks on their towns, and a succession of Indian assaults against Hadley, Hatfield and Northampton did not succeed. The colonists also realized how clumsy they had been when trying to stage offensive actions in the forest, falling time and again into ambuscades.

Captain Benjamin Church insisted that the English needed Indians of their own to serve as guides and scouts, and these were recruited from locals who hadn't joined King Philip's cause. On May 18, Captain William Turner led 150 men against a large encampment near a waterfall on the Connecticut River. He surprised the village and slaughtered over a hundred of its residents, more than a hundred more perishing in their attempts to escape on the river and over the falls. Indian counterattacks killed Turner and 37 of his men; however, the battle still had a demoralizing effect on the tribes, who, for all their success of the previous year, had still not been able to foresee an ultimate victory over their stubborn foe.

The end of King Philip's campaign was hastened when he was attacked from the west by the one military threat that everyone in northeastern America feared: the Mohawks. This branch of the Iroquois Confederacy began ravaging his forces from behind, even as the colonists seemed to gain strength in front. On August 12, 1676, King Philip was surprised and shot while still planning to pursue his offensive with severely depleted forces. With his death, a brief period of relative peace came to the Connecticut River Valley.

The cathartic King Philip's War ended any possibility that the small New England tribes—Wampanoag, Nimtuck, Narragansett, Mahican—could wrest back their beloved land from the Pilgrims and Puritans who had come streaming out of England. The small tribes had not only been defeated but dissipated; the Indians who remained within the borders of New England no longer had to be dealt with as "nations," but solely on an individual basis.

By the summer of 1677, Englishmen had returned to the burned-out site of the village on the Connecticut River, now Deerfield née Potumcuck; nevertheless another, aberrant, Indian raid killed a man and captured three others and a boy. The Indians were remnants of King Philip's men, and their raid illustrated the fact that, whether there was a war on or not, life on the frontier was fraught with peril. These Indians, in fact, had found new leadership, and they marched their captives north to deliver them to the territory that would comprise the next great threat to New England: New France.

It may seem odd that Frenchmen under the Bourbon dynasty—then embodied by its "Sun King," Louis XIV—proved so adept at carving out a bastion of their civilization from the wilds of southern Canada, but their initial success in colonizing the New World was remarkable.

Frenchmen turned out to be not only good Indian fighters, but also astute in establishing Indian alliances. The term *coueurs d'boise* describes these rugged immigrants who divested themselves of European dress, and even customs, and became within a generation or two veritable "natives" of the wilderness. In addition, French explorations deep into the interior had created a geographic framework that threatened to surround the English and those settlers who had preceded them, particularly along the Hudson River, the Dutch. By 1670, in fact, Frenchmen had already traveled down the St. Lawrence River and among the Great Lakes, to the sites of present-day Pittsburgh, Cleveland, Detroit, Chicago and St. Louis, and had sailed the Ohio and Mississippi Rivers to the Gulf.

When the Protestant champion William of Orange became King of England in 1688, the English came in on the side of the European states who were already at war with Louis XIV. In America, New France thus became pitted against New England and its neighbor, albeit separated by fifty miles of dense woods, New York.

At the beginning of King William's War (as it was known in America), the three great powers of the northeast were the English, the French and the Iroquois, especially their fierce, easternmost component, the Mohawks. The New Yorkers had already established excellent trading relations with the Mohawks and had achieved a relationship that amounted to an alliance. The French, on the other hand, benefited from the friendship of not only those Indians opposed to the Iroquois, but those with well-earned animosity toward the English. The French, moreover, outnumbered some seven to one by their colonial enemies, didn't employ Indians merely as scouts or auxiliaries; the French used them as troops. When Indian tactics and battlefield ferocity became allied to French leadership and discipline—and vice versa—the danger for Englishmen on the frontier again increased.

Although the new war did not result in the wholesale devastation of the Connecticut River Valley that King Philip's had wrought, it lasted far longer and Deerfield, once more on the front line, suffered more attacks than any other town. The worst was probably when Thomas Broughton, along with his wife and children, were killed by Indians, and three young daughters of the widow Wells were scalped, two of whom later died. French-led Indian raids were small in size, not more than fifty men to a mission, and basically aimed at keeping New England preocupied with home defense.

The war was fought in New York as well as in New England, Schenectady being pillaged with the death of almost sixty colonists. New York's allies, the Mohawks, had meanwhile launched a 1,500-man invasion that almost reached Montreal, destroying the village of Lachine and killing several dozen French. An English force under William Phips took Port Royal in Nova Scotia—then French Acadia—and tried, but failed, to take Quebec. In New England, the fighting dwindled gradually into ongoing suppressive raids, designed to keep everyone terrorized and on the defensive. The major French expeditions were launched against the Iroquois, their greatest nemesis, and they devastated a number of Seneca, Onondaga and Mohawk villages.

By the time the Treaty of Ryswick was signed, Deerfield had suffered twelve dead, 37 wounded and four captured, but the frontier town had also gotten very good at defending itself. Its sentries were vigilant and its men tough. Best of all, the stockade in the center of the new community had been enlarged and strengthened, and was sufficient to deter all but the most large-scale attack. By the last stages of King William's War, French and Indian raiding parties were bypassing Deerfield to attack the next town down the river, Hadley. The frontline outpost of Deerfield was simply too formidable.

In 1697, after the monarchies of Europe called off their war, an edgy peace settled once more over the English colonies, while to the northwest the French finished off their retributive campaign against the Iroquois. Disease also broke out among the New York tribes and by the end of the century the population of the Iroquois Confederacy had been cut in half. Their English "allies" proved to be no help at all against the French, and the Mohawks sued for peace, not giving up their trade with the English, but no longer remaining hostile to New France. The balance of power had shifted.

Relations between New England settlers and local Indians, which had become difficult enough after King Philip's War, deteriorated further after King William's. The culmination was the edict out of Boston that declared: "Any Indian found within 20 miles of a settlement will be hung." Most of the casualties in Massachusetts had come from surprise, seemingly random, attacks, and even those Indians who had always been friendly to the English came under suspicion of being secret agents of the enemy.

On the other hand, it's a wonder that New Englanders had any Indian allies at all, considering their "scalp policy." Even before King William's War, the government of Massachusetts had offered a bounty of £10 for an Indian's scalp, and during the war this reward was increased to £50. (The price for scalps of small children humanely remained at £10.) Although the Puritan governors of the colony, with their strict moral code and insistence on proper behavior, no doubt assumed their bounties would be claimed only by individuals dedicated to the common good, the policy was barbaric and an open invitation to atrocity. Aside from being an inducement to kill prisoners, one shudders to consider the plight of a young Indian who was accosted at a disadvantage by an Englishman in debt, or even one with an entrepreneurial mind. French missionaries had meanwhile been traveling, at great risk, into Iroquois territory, in attempts to instill Catholicism in the "heathen." Many Indians—Mohawks and others—moved north as wards of New France.

After 1697, there were a few brief years when the frontier was calm, although this is not to say that life in the New World was easy. Even without warfare, the creating of communities and making a living from the timber-covered land was difficult for English settlers, the life a hard one. Government attempted to apply the strictest compliance with Puritan moral conventions, perhaps all the stricter because the only other influences in the New World came from raw nature, the natives, or the papists. In 1692, as if anyone needed a reminder of the many "evils" New Englanders were forced to resist, in the town of Salem a number of settlers, primarily young women, had turned into "witches" and had needed to be put to death.

Whatever challenges were faced by European settlers in the New World, events across the Atlantic still cast a huge shadow, and in 1702 a disaster occurred when Charles II of Spain died without an offspring. Shortly before Charles's death, Louis XIV had convinced him to name his own grandson, Philippe d'Anjou, as the heir to the Spanish throne. This move, patently intolerable to the English, Dutch,

Austrians and everyone else, set off the War of the Spanish Succession, in American shorthand known as "Queen Anne's War."

Although the New World was once again a sideshow as the great European armies, including those led by Marlborough and Villeroi, took the field, a glance at the map will reveal the peril faced by the English in America if France and Spain united. Although the French were established in the north of the continent, with its less-than-desirable climate, their hold on the St. Lawrence River presented them with a gateway to the interior that was denied the English, who butted against not only the Appalachian, Catskill and Adirondack mountains, but also against numerous independent Indian tribes that held the land between.

It was French explorers such as La Salle and Joliet who were "founding" the interior of America, presumably creating alliances with tribes along the way. The Spanish still held Florida, plus Mexico and what is now Texas, and had duplicated the extent of French exploration of America from the south. In addition, both these nations were established in the Caribbean. Though neither of their navies had proven they could stand with "Britannia" in battle, if they were to band together in the New World, in a fleet action, England would be in trouble.

As for the hardy Puritans of New England, patriotism for the mother country was not in short supply, and particularly after they had fought off French-backed Indians for nine years, prior to 1697. On April 24, 1701, Cotton Mather preached, ". . . there must be another Storm, and War, before all clear up, according to our Desires." King Philip's War, the rising of the New England tribes, had been, as a percentage of population, the most costly Indian war in American history. During King William's War, the Indians had found new leaders, and allies, in the French. The French themselves, in unknown strength, lay north of New England, malevolently plotting the destruction of English settlements with their surprise, obliterating attacks that used the manpower and methods of Indian warriors. The English, having detached themselves from reliance on local Indians, were not adverse to a war that would eliminate the French menace once and for all.

Deerfield, Massachusetts, as a frontier town, was not peopled by innocents. As the population of the town grew, by 1704 its residents increasingly consisted of rough-hewn individuals who knew their role, perhaps also attracted by the fact that Deerfield residents paid no taxes. Its population was younger and had a higher average record of crime than that in most towns (although in New England at that time, serious offenses could be anything from laughing in church to wearing silk). A high percentage of Deerfield's residents were also in debt. In any case, the town had held its own during the previous war, and it was a place for young men, or young families, to feel needed. Indeed, as a far-flung outpost of English settlement, because of its strategic importance its population was forbidden to leave.

Indicating the importance of Deerfield to New England as a whole, one of the bright young stars of the colonies was dispatched in 1686 to serve as its pastor. Reverend John Williams, recently graduated from Harvard, and, through both himself and his wife (the former Eunice Mather) related to the leading citizens of Massachusetts, assumed a role that was as much regent as spiritual leader, in that he became the most influential connection between the frontier outpost and the General Court, the government of the colony. The citizens of Deerfield built him a farm, as well as a home within the town's stockade, and voted him a salary of £60 a year.

The town's stockade, over ten feet high, encompassed fifteen houses in all, as well as the big meeting house and commons. It measured about 200 by 300 yards, and had three gates, "strong and substantiall, with Conveniencyes for fastning both open and shut." The home of Thomas Wells, an officer in the local militia, which stood just south of the fort, was also fortified as a "garrison house," suitable for all-round defense. Since Deerfield was a farming community, most homes were situated outside the stockade; however, the fort was built to provide accommodation for all 270 Deerfielders in an emergency or a time of crisis. During wartime, an enemy raid could occur at any moment and would always be unexpected.

As during King William's War, the French never entertained the prospect of conquering New England. They simply wanted to keep it on the defensive with random attacks. New York remained tacitly neutral at the beginning of Queen Anne's War. Having subdued New York's allies, the Iroquois, the French no longer had no interest in attacking New York, and may even have promised the Mohawks they would not do so. The French subsequently appreciated that colony's subsequent lack of motivation to fight New France. New England, however, was another matter, and the French found ready allies who were eager to assist in raids against the arrogant Puritans and Pilgrims of Massachusetts, Connecticut and Plymouth.

In the south, Carolinians welcomed the war with an attack against Spanish St. Augustine in 1702. In 1703 they launched an attack against Pensacola. In the north, even as New Yorkers continued their commerce with Montreal, both New France and New England attempted to cultivate the strong Abenaki tribe of Maine. On June 30, 1703, however, the French, using some Abenaki warriors along with their "French" Mohawks, attacked two Maine towns, Saco and Wells. New England responded the only way it knew how: retaliating against the nearest Abenaki villages. The entire Abenaki tribe now came in with the French.

In Deerfield, the frontier outpost closest to New France, rumors of impending French and Indian raids came in steadily throughout the summer and fall of 1703, putting the town on edge and forcing a defensive posture. On October 8, in fact, Indians surprised and captured Zebediah Williams and John Nims just outside of town and took them to Canada. Reverend Williams wrote to the governor of the colony, Joseph Dudley, in Boston, "We have been driven from our houses and home lots into the fort . . . the frontier difficulties of a place so remote from others and so exposed as ours are more than can be known . . ." The Boston government said it had 1,900 armed men along the frontier—mostly citizens of the towns themselves, but also some regular troops. On February 24, 1704, twenty soldiers of the colony marched into Deerfield and were put up in scattered houses within the powerful stockade.

In Quebec, Governer Philippe Vandreuil orchestrated the French terror offensive, and in the deepest winter of early 1704 decided, apparently merely for the sake of wreaking unexpected havoc, to launch a larger-than-usual raid against a New England town on the opposite side of the frontier from Saco and Wells.

A party of about fifty French regulars and militiamen, under the command of a young officer, Hertel de Rouville, was dispatched south, together with 200 Indian warriors. The Indians were primarily Abenaki, although some were Catholicized Mohawks, along with a scattering of others with an earnestness to fight the Puritans. Heavily armed and clothed, wearing snowshoes, the expedition journeyed south in early February 1704. Easily skimming across frozen rivers and lakes, the party left its sleds and dogs at a spot near present-day Brattleboro, Vermont. Then they marched, with greater caution, another twenty miles until, shortly after midnight, they were on the outskirts of Deerfield.

As terrifying as French and Indian raids could be, months passed, even in a frontier town, without seeing signs of the enemy, and vigilance sometimes relaxed. In the brutal cold of winter, and surrounded by fields of snow, everyday life could sometimes seem a greater imperative than military defense. In the dark, early-morning hours of February 29, 1704, most Deerfielders were nevertheless huddled within their uncomfortable accommodations in the fort. Some daring families, taking a gamble with fate, remained in their homesteads, outside the protection of the stockade. Within the fort, a sentry was on duty throughout the night, supposedly armed and ready to give warning, should an enemy appear. In the silent, early hours of February 29, in fact, at least 250 French and Indians were less than a mile from the town—and approaching.

With guns primed, and knives and hatchets at the ready, the marauders advanced slowly across the field in the dark, stopping every few moments to let the sound of their feet on the snow dissipate innocuously in the wind. They passed the empty homes at the north of the town; still, not a sound could be heard from the black, silent fort. Then, to their incredulity, they saw that snow had piled up against the town's stockade. It made a ramp they could use to easily surmount the walls.

Some two dozen Indians quietly snuck up and hopped into the fort, then opened the nearest gate for two hundred others to creep in. No one has recorded how many minutes the French and Indians took to spread throughout the interior of the fort before the sleeping sentry woke up and fired a shot, shouting "Arm!" just before his death. By that time it was too late for the English.

The raiders were able to rush straight into some Deerfield homes; in others they had to chop their way in. Englishmen in sleeping clothes grabbed weapons and rushed to protect their families. Some dashed outside their homes, not realizing the enemy was already within the stockade. Gunfire, screams and war-whoops resounded throughout the fort. The enemy was intent on taking prisoners, but after their excruciating 300-mile journey from New France, they were not inclined, at this moment of encounter, to be kind. Babies were automatically killed, since they were bad captives; fighting men were triumphantly slain. Younger women were good prospects as captives and teenaged children excellent.

Reverend John Williams, in his book *The Redeemed Captive Returning to Zion* wrote: ". . . before the break of day, the enemy came in like a flood upon us; our watch being unfaithful . . . I leapt out of bed, and running towards the door perceived the enemy making their entrance into the house; I called to awaken two soldiers in the chamber, and returned to my bedside for my arms; the enemy immediately brake into the room. I judge to the number of twenty, with painted faces and hideous exclamations. I reached up my hands . . . for my pistol, uttering a short petition to God for everlasting mercies for me and mine. . . . I cocked it, and put it to the breast of the first Indian that came up, but my pistol missing fire, I was seized by three Indians who disarmed me." Williams watched while the Indians killed his baby daughter, Jerusha, his six-year-old son, John Jr., and his black servant, Parthena.

The French troops, in the meantime, ensured that the victory would be total. Any homes the Indians had neglected to torch were set afire, and pockets of resistance systematically wiped out. Unfortunately, Sergeant Stebbins' house, in the northwest corner of the fort, could not be entered. The Stebbins house had seven men in it, including three garrison troops, and brick-reinforced walls. According to a contemporary account, the house was "attaqued later than some others," and they had some minutes to get ready for the onslaught. They stood "stoutly to their Armes, firing upon the Enemy and the enemy upon them . . . with more than ordinary Couridge." A French officer was killed assaulting this bastion, as well as "three or four Indians," one of them a chief. John Williams recorded, "The judgement of God did not long slumber against one of the three [Indians] which took me, who was a captain, for by sun-rising he received a mortal shot from my next neighbour's house."

Next to Stebbins', John Sheldon's house also held out, although an Indian, after hacking a hole through the door, was able to get off a shot that killed Mrs. Sheldon, who was in her bedroom. (The door, tomahawk marks and all, is currently on display at the museum in Deerfield.)

In other homes, settlers had taken shelter in their cellars, only to die when their houses were put to the torch above. At least one entire family—John Hawks, his wife and three children—succumbed together in their cellar, asphixiated. Others found this a successful maneuver when the fires didn't catch. One woman hid under an overturned tub; a boy hid under a pile of flax. A young couple and their infant, living in one of the "small houses" in the fort, escaped detection because their home was completely covered by snow. One man was captured, and had his right forefinger cut off, per Iroquois custom, but then escaped. Mrs. Mary Catlin tended to a wounded Frenchman who was carried into her house. Her entire family was killed or taken, but the enemy left Mary behind.

Some English had been able to escape over their roofs and the stockade, including six men who left their families so that they could assemble at Captain Wells's garrison house, just outside the fort, for a counterattack. These men's families suffered disproportionate casualties inside the fortress/deathtrap. Outside the stockade, and particularly to the south, the settlers were unmolested. One young boy, John Sheldon, Jr., had hopped out of the fort at the very beginning and, stripping off pieces of his shirt to bind his bare feet, had run across the snow to the nearest town, Hadley. There, the glow of fire in the sky was already visible, and John Jr.'s report of what was taking place caused much excitement.

Inside the fortress of Deerfield, the French and Indians assembled their valuable captives, setting fires, hunting down fugitives, and slaughtering lambs, cows and pigs to add to the general carnage. It was over an hour past daylight and some of the force had gone off with their captives while others mopped up the pitiful, beaten remnants of the town. Prisoners were compelled to dress themselves warmly for a long, winter journey.

During this stage, when the victors possessed such a large number of prisoners of all sizes, gender and age, one false move by a Deerfielder could result in immediate death. Rev. Williams recorded that "Indians insulted over me for a while, holding up hatchets at my head." As the captives dressed themselves while at the mercy of their attackers, anyone who revealed a look in their eye conveying either excessive truculence or weakness would be killed rather than taken. One can only imagine the stoicism of the captives, mixed with the sounds of sobbing.

Suddenly, yells and shots were heard, as well as the sound of horses' hooves pounding across the snowy landscape. The French and Indians were jerked from their triumphal reverie to glimpse the onset of their greatest fear: an English counterattack bearing down on them from the south.

In fact, it was a mere thirty-five armed men from Hadley and Hatfield, who had rushed to the battle, after being joined by fifteen unscathed Deerfielders who were as yet unhurt south of the stockade. Nevertheless, the enemy had shot his bolt in the hideous sacking of the sleepy town, and fifty fresh, well-armed Englishmen attacking with revenge in their hearts was an unwelcome development.

The French and Indians ran for their lives, abandoning loot and some of their captives. The English charged into the fort through one gate; the enemy ran out the opposite. The avengers pursued the invaders across the snowfield toward the river, "killing and wounding many," meantime throwing off their heavy clothing in their enthusiasm to seize back victory, and the captives. This is when the factor of Indians led by French had another impact on the battle.

The English leader of the counterattack, Captain Wells, ordered his men to hold up; the village had been resecured, but now the counterattack was becoming strung out, no longer coordinated. His men did not listen and charged all the way to the Connecticut River. Then they ran into a wall of fire.

Hertel de Rouville, with his force of over 200, was not ready to be stampeded by fifty English, and those of his men who had already vacated Deerfield simply lay in wait, in ambush. When the overly enthusiastic rescue party, whose "courage was more worthy of applaus than their conduct," approached the river, a line of French and Indians suddenly rose from behind the banks and fired. Then they charged. The Englishmen fell back, "facing and firing, so that those who first failed might be defended." In the fight on the meadow between the town and river, nine English were killed, several others wounded. The survivors got back to what was by then, once more, the "safety" of the burning, blood-drenched stockade.

Now the French and Indians could calmly organize their march north, with their 109 captives. They proceeded at once. The English stood behind their walls, shaken and stricken with grief. Deerfield had been, for all practical purposes, destroyed: 47 dead within the fort, mostly women and children; others wounded, plus the nine bodies of the valiant counterattackers out on the field. Of the twenty soldiers who had arrived just days before, five were killed and five captured. New England had suffered a defeat, early in Queen Anne's War, that would echo morbidly throughout the colonies.

French and Indian losses are more difficult to assess: Governor Philippe Vandreuil stated in his report that only eleven men were killed in the victory. John Williams, though, after residing in New France reported, "After my arrival in Quebec, I spake with an Englishman, who was taken the last war, and married there and of their religion; who told me, they lost above forty, and that many were wounded. I replied, the Governor of Canada said they lost but eleven men. He answered, it is true, that there were but eleven killed outright at the taking of the fort, but that many others were wounded, among whom was the ensign of the French, but (said he), they had a fight in the meadow, and that in both engagements they lost more than forty. Some of the soldiers, both French and Indians then present, told me (said he), adding that the French always endeavor to conceal the number of their slain."

What Williams heard corresponded to the eyewitness reports of Jonathan Wells and Ebeneezer Wright, who had fought in the English counterattack. They said they saw ". . . many dead bodies and . . . manifest prints in the snow" to indicate the enemy had dragged off their dead. They estimated "fifty men [killed] and twelve or fifteen wounded," who will "not see Canada again." Williams, then a captive, also mentioned that, "after this fight, I saw no great insulting mirth, as I expected; and saw many wounded persons, and for several days together they buried of their party . . ."

Around the town of Deerfield, within a day, 250 men were assembled to pursue the enemy and retrieve the captives. However, they lacked snowshoes, and the weather had warmed, making difficult slush. And besides, it was not clear the French and Indians could be overtaken, given their head-start. There was no pursuit.

As for the Deerfield captives, twenty-one of them were murdered on the march north, in almost every case small children or women who couldn't keep up. Mrs. John Williams, who had given birth to Jerusha the month before, tripped as she was crossing a stream, and her Indian captor, disillusioned at her prospects of continuing in frozen clothes—or perhaps her fall was the final straw—put her to death with a hatchet-blow to the head. An eleven-year-old girl was also killed that day, as well as an infant. Reverend Williams lamented, "I was made to mourn at the consideration of my flock's being so far a flock of slaughter."

On the fourth day of the raiding party's retreat, another woman was killed, and then four more on the fifth. Two additional were put to death by their captors on the seventh day, and on the eighth a woman named Mary Brooks approached the Reverend. She reported that she had fallen down several times and had suffered a miscarriage. "I know they will kill me today," she said. "I am not afraid of death . . . Pray for me at parting, that God would take me to himself." She was, indeed, put to death on the eighth day of the journey to New France.

There has been a trend among recent writers on the Deerfield battle to view the Indians' killing of prisoners with astonishing equanimity. After all, executions like that of Mrs. Williams could be accomplished "at one stroke," as opposed to allowing the victim to stagger on needlessly, perhaps become prey for animals and, most important, impede the progress of the raiding party. (Perhaps, in future centuries, the actions of Japanese officers at the Bataan Death March will come to similarly be "understood.") It is beyond question that colonial-era warfare was brutal, villages and towns on every side being the most frequent objectives, civilian scalps and prisoners the "prizes." However, for twentieth-century chroniclers to rationalize noncombatant murder is inconducive to understanding that terror was a primary weapon in those early wars for America. Warmaking against civilians was (and still is) a horror.

In any case, that Mrs. Williams' body was recovered a few hours later, and brought back to Deerfield for burial, indicates that there were at least some English "Hawkeyes" trailing the enemy retreat through the forest in its early stages, even as the major force declined to pursue. Regarding the bodies of the others, there is no record.

The French parted ways with their allies after reuniting with their sleds and dogs. They hastened back to New France via the most direct route, where De Rouville basked in the glow of recognition of a job well done. John Williams recorded, about the march north, that "Indians alone are responsible for the Deerfield contingent; the French take no part in captivity." Nevertheless, once Williams reached New France, he said, "The French were very kind to me. A gentleman of the place took me into his house and to his table, and lodged me at night on a good feather bed."

The Indians made some good money for their captives, significantly in contrast to what they could have made for their scalps. Governor Vandrueil reiterated his policy to the Abenaki in 1706: "Since the beginning of this war I declared to you that this manner of paying for scalps seemed too inhuman, but that I would give you ten Spanish crowns for each prisoner." In fact, depending on the prisoner, the Indians often received more, which also explains why the death toll at Deerfield wasn't higher than it was. Back in New England, meanwhile, the price for Indian scalps during Queen Anne's War shot up to £100 apiece.

The French were also different from the English in their treatment of captives, and many Deerfielders found themselves with absolute freedom within the French colony, whether held in Indian villages or small settlements of *habitants*. (The English generally kept French captives in prison.) One young Deerfielder ended up with a good business around Montreal, and took part in explorations of far-flung French territory, going all the way to the Mississippi. Some captives, including John Williams' seven-year-old daughter Eunice, became "Indian," and never did return to English society.

In the spring following Deerfield, New England put together a major land and naval force that rampaged up the coast of Maine, destroying Abenaki villages, although some of these had been abandoned. Benjamin Church, the hero of King Philip's War, provided good leadership, although he was by now so obese that he needed a soldier to march alongside, to help him over fallen logs. Then, in 1705 and 1706, the war entered a lull, during which prisoners were exchanged, including Reverend Williams, who returned to Boston to much rejoicing.

As for the remainder of Queen Anne's War, the French and Spanish raided Charleston, South Carolina in 1706, and then English and Indians raided Pensacola twice in 1707. The second time, the

Spanish were relieved by French out of Mobile, coming to their rescue with (Alabama) Indians of their own. Two attempts were made against France's Port Royal in the north in 1707, but then French and Indians hit Haverill, just north of Boston, in 1708.

In 1709, a huge expedition was mounted against Quebec, but was then called off. Port Royal finally fell the following year, and in 1711 an even larger army was launched against Quebec. Now that British regulars and components of the fleet were finally on hand, even the Iroquois joined in. The expedition met with a maritime disaster in the St. Lawrence, however, and was then discontinued; the Iroquois promptly resumed their neutral stance toward New France. In 1713, the European powers signed the Treaty of Utrecht, and the second war between the French and English in America ended.

The third, King George's War in 1739, saw the English capturing the fortress of Louisbourg, while French sacked Saratoga, New York, taking over a hundred prisoners. This conflict was finally called off in 1748, with terms that required Louisbourg to be given back to the French.

The final French and Indian war, called just that in American history (the Seven Years' War elsewhere), broke out in 1755. By this time the number of English colonists had grown considerably and British regulars were also involved from the start. A thousand of these under General Edward Braddock, along with several hundred colonial auxiliaries, marched on French Fort Duquesne near present-day Pittsburgh, only to be disastrously defeated by French and Indians in an improvised encirclement battle in dense forest. It was not until British General James Wolfe, with an army of regulars, appeared in the early morning hours on a field outside Quebec in 1759 that the center of French power in America was finally crushed. The taking of Quebec was a triumph for British arms; however, it was to be one of their last on the American continent.

After a century and a half of struggle, mostly without any assistance at all from the mother country, and including such traumas as took place in Deerfield, Massachusetts, the English who settled the New World had begun increasingly to feel a sense of "independence." The British would soon have to fight another major war in North America, starting in 1775, but this time it would be against their own American colonies. The ultimate American victory was won largely because of an army, and a fleet, that was sent to their assistance from France. Twenty years later, Thomas Jefferson paid $15 million to France for the remainder of its American territory, in the Louisiana Purchase.

3

John Paul Jones off Britain

BY JAMES M. ALDRICH

The sailor's bare feet slipped on the damp cobblestones. He was a sailor—his tarred pigtail and rolling gait brooked no doubt. Even though darkness hid him, he glanced often behind, almost as if the devil himself pursued. He knocked at the first house he encountered. There was no answer. By the third house, he was yelling as he pounded on the door. When the owner at last unbarred it, the breathless sailor could only point back towards the harbor, where fire blossomed, and a ship burned.

This sailor was a deserter from the Continental Navy sloop *Ranger*. His captain, John Paul Jones, was busily reminding the English people of a watery fact: The Atlantic and the Channel may well have been English moats, but they were also highways to the English heart for any raider willing to "go in harm's way."

After the thirteen American colonies declared independence in 1776, Great Britain, as well as her numerous European enemies, began to realize that this was serious business—a war for separation and not just another outburst from some fractious colonies. By the close of 1777 the former colonists, fighting on their own turf in their own fashion, had demonstrated at Bunker Hill, Trenton, Princeton, and particularly at Saratoga, that they could at least hold their own against British land forces. Their prospects for success at sea, however, were quite another matter.

Mighty Britannia did indeed rule the waves, but her navy was sore-pressed to maintain that rule. With Britain confronted by an increasingly hostile Europe led by France, and serious problems in the

Caribbean and on trade routes elsewhere, some leading Americans saw an opportunity for success at sea. Two things were needed: ships, and substantial foreign support, preferably from France. Experienced seamen and leaders of courage and determination were already on hand.

To meet the need for ships, the Continental Congress undertook a shipbuilding and ship borrowing/purchasing program. To pursue foreign help, three colonial leaders were sent to France as "Commissioners," with Ben Franklin as their leader. By the close of 1777, acquisition of shipping and negotiations securing France as an ally were both nearing fruition. Still, the British naval juggernaut could not be directly challenged on the seas. So the fledgling American Navy adopted a policy of commerce raiding, hoping that their opponent could be bled to death as surely by a thousand cuts as by a single wound. If those raids extended into British home waters, then national prestige as well as commerce would be injured.

Enter John Paul Jones, described by various historians as disputatious, insufferably cocky, a superlative seaman, a fighter, a leader of genius, an accomplished officer, and an overproud Scot. Born John Paul, son of John Paul, in Kirkbean, Scotland on the shores of the Solway Firth, he had entered this world on July 6, 1747. The senior John Paul was a landscape gardener about whom little is known.

Young John Paul (he was never referred to as "Junior") had an elder brother, William, who had migrated to Fredericksburg in the American colony of Virginia. By the time young John first went to sea at age 12, William was already established as a tailor there.

It was in 1759 that John Paul was apprenticed to a Scots merchant-shipper named John Younger. He sailed that year as a cabin boy on a voyage to Virginia, where he visited his brother. Until 1766 he continued in service with Younger, making several voyages to the American Colonies. Then Younger's business failed and the apprenticeship ended.

Shortly thereafter, John Paul received a warrant as midshipman in the Royal Navy, but he never served (fortunately for the United States!). Records relate that in 1768 he engaged in the slave trade as mate aboard a Jamaican brigantine. Accounts vary, but for several years he continued in that grim employment until taking passage to Scotland in a brigantine. Both the master and the first mate died of fever during the voyage, and John Paul took over command, bringing the ship safely into port. For this accomplishment he was appointed master of the vessel, having obtained his master's warrant a year or two earlier.

John Paul purchased a ship of his own in 1772. The following year he dealt with a mutiny off Tobago by killing the ringleader. Taking the advice of friends, he fled to America prior to the convening of the Admiralty Court. Though their sequence remains unclear, three significant events changed his life over the next two years.

John Paul inherited part of the estate of his brother in Virginia, which encouraged him to remain a colonist. He changed his surname to Jones, perhaps to stay a step ahead of a murder charge. Finally, in the autumn of 1775, he received a commission as the senior lieutenant in the new Continental Navy.

The following year he rose to command of the sloop-of-war *Providence*, in which he patrolled North American waters seeking out British shipping. The end of this cruise showed the remarkable tally of sixteen prizes taken. More important, his successful raid on the undefended anchorage of Canso Bay, Nova Scotia planted the germ of an idea that would eventually take him to a distant English shore.

Perhaps no less have been expected from an officer who wrote, "I wish to have no Connection with any Ship that does not sail fast, for I intend to go in harm's way . . . ." The Continental Congress recognized Jones's achievements with promotion and assignment to the command of the best fighting ship in the young navy, the newly built sloop-of-war *Ranger*. With its armament of 18 six-pounders, the sloop was powerful for its class, and destined for great things under its new master.

On November 1, 1777, Jones sailed for France carrying word of Burgoyne's surrender at Saratoga. His orders were to report to the American Commissioners, from whom he received permission for an offensive cruise around the British Isles. After a short period of refitting in French ports as he was preparing for his cruise, John Paul Jones had the honor of receiving the first foreign salute to the newly adopted American flag—by the French on February 14, 1778, in Quiberon Bay.

By the end of March all was ready, and *Ranger* made preparations to enter the English Channel. There is little doubt that Jones had a firm plan, perhaps partially based on these words written to him in the previous year by Robert Morris: "It has long been clear to me that our Infant fleet cannot protect our own Coasts; and the only effectual relief it can afford us is to attack the enemies' defenseless places and thereby oblige them to station more of their Ships in their own Countries, or to keep them employed in following ours, and either way we are relieved." American citizens had long trembled at the sound of British boots. It was past time for British citizens to learn the meaning of fear.

Aside from the fear factor engendered by raiding a British city, Jones also planned to abduct a key British official and exchange him for the hundreds of American prisoners rotting in British prison hulks. His time as a slaver had taught him the horror of such involuntary confinement, and engendered an appreciation of liberty. "When I entered into the service," he later wrote, "I was not actuated by motives of self-interest. I stepped forth as a free citizen of the world, in defense of the violated rights of mankind. . . ."

John Paul Jones' final plan was a raider's dream, and a defending admiral's nightmare. With one small vessel, he would infiltrate the enemy's Channel patrols and enter the Irish Sea. While destroying merchant shipping left and right, Jones planned to raid at least one English city, then abduct Lord Selkirk from his home on St. Mary's Island in Solway Firth. Finally, he would return to France, again running the patrol gauntlet while taking prizes.

Yet, he did not start without a major handicap. The crew of *Ranger* had been long away from home. Their pay was in arrears, and their captain had driven them hard to ready their vessel for the raid. When several men deserted for local French fleshpots or a chance merchantman for home, it was no surprise to Jones. That the discontent ran so deeply through officers and men alike, however, exploded as a near-deadly surprise as the raid progressed.

On April 10, 1778 *Ranger* sailed, escorted for the next four days by a French frigate. Even as the frigate parted company, Jones captured and burned a brigantine loaded with flaxseed. On April 17, he captured the ship *Lord Chatham*, loaded with porter. A prize crew delivered the vessel to Brest, in France, five days later. The next day, *Ranger* was challenged by the revenue cutter *Hussar*, which managed to escape capture only by its superior sailing qualities (it could sail closer to the wind due to fore-and-aft rig). On April 19, a schooner with grain aboard was burned, an armed cutter escaped pursuit, and a sloop in ballast was set afire. The next day, a fishing vessel provided information that the Royal Navy sloop *Drake* was anchored in Belfast Lough. Jones wanted to sail in immediately and "cut her out." His crew—their discontent increasing proportionally to the danger— refused, although they agreed to try during the night. The first attempt to anchor off the *Drake*'s bows failed because of drunkenness aboard the *Ranger*. Though the *Drake* was not alerted, worsening weather prevented a second attempt, and Jones barely made it to the open sea without wrecking.

On April 22, Jones decided that the time was ripe to raid an English port. Information that he was in the area, visible in columns of smoke and carried by the escaped cutters, would soon have the Royal Navy around his ears. That he picked Whitehaven—the port from which he had first sailed to America—as his target was neither ironic nor any measure of vengeance. Rather, cold logic dictated that the best target was one that Jones knew well, and a port sheltering large quantities of shipping. Whitehaven simply fit the bill. John Paul Jones stood ready to beard the lion in its den—a goal no raider had achieved since the Anglo-Dutch wars over a century earlier. Unfortunately, his crew suffered an extreme lack of enthusiasm.

Not only were they far from home, but they had signed on for prize money. Their captain had burned their prizes, and now proposed to burn a harbor full of them! Disgruntlement turned to anger, and anger to mutiny. Even the first and second officers were involved. Fortunately, the third lieutenant warned Jones. When he was rushed by the crew in the early-evening hours, they found him with a cocked pistol. Little is recorded of that dark hour, but somehow Jones managed to restore order without bloodshed or a single man being placed in irons.

The ship sailed steadily towards Whitehaven—until the wind died, miles from the harbor. Manning two ship's boats (estimates of the number of men in the boats varies from thirty-one to forty), sometime around midnight Jones began what would be a hard three-hour pull to the harbor. Few raids have begun so inauspiciously. Bedeviled by mutiny and wind, who could have blamed Jones for surrendering to the inevitable? If he reached the port and if he returned, there was no guarantee that his mutinous lieutenants would not have fled with the *Ranger*. Yet Jones continued. That stubborn streak in John Paul Jones, that determination to succeed at any cost, remains the hallmark of the successful raider.

The boats separated at the entrance to the harbor. Jones rowed to spike the guns in the harbor batteries, while his lieutenant of marines led the remaining crew to fire the shipping that crowded the port. Jones's boat grounded below the south battery, unobserved. Leaving his loyal third lieutenant to guard the craft, he and the remaining men ran to the emplacement walls. Lacking grappling hooks, they climbed by the simple expedient of forming a human ladder. All was quiet inside; no guards had been posted! The garrison was soon barricaded into its own guardroom, and the men quickly spiked the position's guns.

Perhaps Jones looked to the harbor at this time, puzzling as to why no ships were burning. Were his men waiting for some signal? Were they lost? Had their combustibles failed? Or were they even now prisoners, with militia on the way to take his party? With those worries in mind, he sent all but a single midshipman to wait on the beach, and ran to spike the northern battery, ensuring a path of retreat. The light of dawn was noticeable as the midshipman pounded the last spike home. Possibly, Jones observed the gathering of crowds, aroused by a deserter, as he and the midshipman sprinted to the boat. Leaping in, he ordered the surly crew to pull for the harbor's wharf, as his faithful lieutenant whispered that the men had indeed planned to abandon Jones to the mercy of the English, had he not been on guard.

At the wharf, they were met by the crew of the other boat. Their combustibles were soaked, and they had been unable to fire a ship. It was little wonder that the combustibles had been soaked during the three-hour pull to the harbor. But it was also little wonder that they had fired no ships in the two hours just elapsed, for they had been enjoying themselves in a local grog shop the entire time! Except, of course, for the deserter, who had slipped away to alert the town's populace of the danger to the over two hundred vessels in their harbor.

By five o'clock, bemused and angry locals were beginning to crowd the quay. It was time for the Americans to cut their losses and leave. But Jones saw it otherwise. He had planned to burn ships, and burn ships he would. After setting a perimeter, he stole a light from a local home, then fired the collier *Thompson*. With luck, the fire would spread to nearby ships. Once the vessel blazed merrily, the Americans tumbled into their boats and rowed for the *Ranger*, which had found the wind and was then entering the harbor. As they fled, at least one unspiked gun fired at them, to no avail. By six o'clock, the raid was over. The inhabitants of Whitehaven, assisted by rain, had doused the fire, the *Ranger* was sailing away, and no blood had been shed on either side.

A tired John Paul Jones must have been exceedingly frustrated that morning. He had accomplished virtually nothing, and had lost a man in the raid—his log book listed the deserter as a presumed prisoner. Blaming it on the late start and the reluctance of his men to burn British ships sounded like an excuse. The next step of the overall plan, the seizure of Lord Selkirk to exchange for prisoners, seemed to be the difference between success and failure of the mission.

By afternoon, the *Ranger* anchored off St. Mary's Isle. Jones, with two officers and several men, rowed ashore. Sending word ahead that they were a press gang, the officers and a squad of sailors headed for the Selkirk residence. They were unimpeded, as every man of serving age fled the supposed press!

Upon arriving at the estate, Jones was shattered to discover that Selkirk was away. He ordered his men to return to the ship, but his officers demurred. The British had looted and burned homes in the colonies, and it was time to pay them back. Besides, the men deserved the loot, since they had not been paid and had little to show for prize money. Fearing another mutiny, Jones allowed them to take the Selkirk family silver and nothing else. It was a mark of the man that he eventually purchased the silver and returned it to the Selkirk family.

As the *Ranger* sailed out of Solway Firth, several vessels were burned or scuttled, one prize was dispatched to Brest, and course was set westward for the northern coast of Ireland. All things considered, it had been less than a successful day for John Paul Jones. Frustrated in his primary objectives, with alarm rapidly spreading across the British countryside and along the coast, it was time to sail for France. Only one unfinished piece of business remained.

The next afternoon, April 24, while cruising along the Irish coast in search of prey, Jones found his unfinished business: The Royal Navy sloop-of-war *Drake* came out of harbor to meet *Ranger* off Carrick-fergus, near the city of Belfast. The adversaries were about equal in size and armament, but *Drake* carried a much larger crew, which meant that trying to board it would be disastrous for the Americans. Jones concentrated his fire on his enemy's spars and rigging.

After nearly an hour of furious action, it was clear that the American tactics and gunnery were overwhelming *Drake*. As dusk approached and British casualties became crippling, *Drake*'s second lieutenant surrendered to John Paul Jones. His captain lay near death and the first lieutenant was mortally wounded. *Drake* had suffered 42 casualties, about thirty percent of her total compliment. *Ranger*'s log listed only eight casualties. Both ships had to be almost completely re-rigged before crossing the Channel.

After only twenty-two days at sea, Jones returned to Brest. The capture of a British warship, the return of two prizes to port with the destruction of several others, and the teasing of the English Lion at Whitehaven brought instant notoriety to the young Captain. He was on his way to being a "darling" of the French. His notoriety in England took a different turn, as a 1778 newspaper wrote, "The Captain of the *Ranger*, John Paul . . . some time ago . . . stood a trial in London for the murder of his carpenter, and was found guilty, but made his escape." Not exactly true, but then . . .

The most important aspect of the raid was the reaction of Britain and the British people. Its coastal settlements were in an uproar, as the *Morning Chronicle* and *London Advertiser* wrote: "The ruinous state of our fortifications of many of our sea-port towns, as like wise the open and defenseless posture of many others, at present seems to suggest some very alarming reflections. . . ." Economically, the damage was more real and immediate, as the *Gazetteer* and *New Daily Advertiser* noted: "Such a damp on commerce has the American privateer called the *Ranger* made, that yesterday insurances to Ireland were five guineas per cent that lately were done at one and a quarter."

American morale rose, as well as the morale of their new French allies. British morale dropped correspondingly, as they discovered the vulnerability of their home ports to enemy vessels. Effort and money that would have bolstered their economy was turned to home defense—fortifications and militia. The Royal Navy, already stretched to its limits, was forced to increase its presence in home waters—a fact that the French soon took advantage of. Finally, the rising cost of marine insurance was no trivial matter for a maritime nation. All this because of one small warship and a captain unafraid to go "in harm's way."

Following this successful raiding cruise and the infliction of a humiliating defeat on a British warship, Jones and his crew were treated to an heroic welcome as they returned to resupply and repair battle damage in France. To his surprise and great annoyance, however, John Paul Jones was relieved of his command of the *Ranger*. For several months he was left ashore to fuss and fume while a larger and more important command was being organized for him.

One report suggested that relieving him of command was largely at the insistence of two of the American Commissioners in Paris, who disliked Jones intensely. A different report held that it was the result of a request, perhaps even a demand, by the French Government, whose desire was to capitalize on *Ranger*'s success with a larger and more powerful raid. Twisting the Lion's tail was a popular sport throughout Europe, with the French leading the charge. *Ranger*'s raid under the driving and talented leadership of Jones suggested that perhaps the British Lion was aging, had a few teeth missing, and was vulnerable.

By August of 1779, an expedition had taken shape. History differs on the number of vessels in the squadron, recording between five and seven. The difference may have been that initially two or three privateers were part of the expedition, but soon sailed off to find profit rather than glory. One of these may well have been the large cutter *Cerf*, of 18 guns.

Of the five clearly identified ships, first and foremost was *Bon Homme Richard* (the former French East Indiaman *Duc de Duras*), rated as a 42-gun frigate. It was an older vessel with some rot in its timbers and a leak here and there, although apparently still considered to have a good turn of speed (the important factor to its new captain, John Paul Jones). The vessel's crew of 380 included eleven different nationalities, although the officers were almost entirely Americans. Jones's first lieutenant was Richard Dale, who distinguished himself on this cruise and in a long career in the U.S. Navy. Another lieutenant under Jones was Samuel Nicholson, who later became the first captain of "Old Ironsides," the U.S.S. *Constitution*. Interesting to note, one of the ship's gunners appears to have been a Narragansett Indian.

The second ship was a newly built Continental Navy frigate under a French captain, Peter Landais. Reports on its armament varied from 30 to 36 guns. Named *Alliance*, doubtless to acknowledge the support of France in the war, its captain was to make a mockery of the name throughout this historic cruise. A third frigate in the raiding fleet was *Pallas*, another old French merchantman mounting 30 or 32 guns. It played a significant part in the vicious, bloody fighting to come.

Two smaller French vessels completed the fleet: *Vengeance*, a 12-gun brig (or brigantine), and *Ariel*, described as a small ship of 20 guns. Both were sailing under the American flag, as all French officers of the squadron accepted temporary commissions in the Continental Navy.

On August 14, 1779, the small squadron under the command of Jones sailed from L'Orient, in northwestern France. Cruising the coast of Ireland and then rounding Scotland, a considerable number of prizes were taken. Jones planned to raid the Scottish port of Leith, but the winds turned against him and he sailed on, leaving the neighboring city of Edinburgh in a state of consternation. He also made threatening gestures toward Newcastle-on-Tyne, but continued in a southerly direction, again due to the fickle wind.

During this otherwise successful and productive part of the cruise John Paul Jones appears to have had constant problems with Captain Landais of *Alliance*, who seemed to enjoy ignoring or misinterpreting orders from his commander. His attitude was to prove costly in the very near future.

Still, off the northeast coast of England on the twenty-third of September, the juiciest prize yet hove into sight—a group of merchant ships from the Baltic escorted by two of His Majesty's warships: the 44-gun *Serapis*, commanded by Captain Richard Pearson, and the 20-gun sloop-of-war *Countess of Scarborough*. The fleets met within sight of shore just off Flamborough Head on England's central eastern shore.

Quickly, Jones cleared for action, sent his French Marine sharpshooters into *Bon Homme Richard*'s fighting tops, and made for *Serapis*. *Alliance* stood off at a safe distance, but *Pallas* took on *Countess of Scarborough* and eventually captured the vessel. It was just about dusk when the battle began, but apparently there was ample moonlight, to the extent that crowds began to collect along the high ground on the coast.

As Jones came to grips with *Serapis*, his first broadside was nearly his last. Two of his three heavy guns, 18-pounders in the lower gun deck, exploded, killing the gun crews and causing considerable damage to the deck above. It was decided to abandon the third 18-pounder and sail alongside the enemy to battle them hand-to-hand. For three hours the two frigates pounded each other, the French Marines slaughtering the English deck crews.

Eventually *Alliance* showed up to help, positioned herself across the bow of *Serapis*, and loosed a vicious raking fire. Regrettably, some of this fire also hit *Bon Homme Richard*, but it did help convince Captain Pearson that Jones really meant it when he bellowed, "I have not yet begun to fight!" A grenade from above, tossed into an open hatch on *Serapis*, set off a number of explosions below decks and fired the ship. That did it. The British surrendered.

Only steady work at the pumps had kept *Bon Homme Richard* afloat, so Jones took his wounded and surviving crew members aboard *Serapis*, cut his flagship loose, and let it sink. The fires aboard the captured British vessel were extinguished, immediate repairs completed, and the squadron sailed for the neutral Dutch roadstead of the Texel, arriving there safely on October 3, 1779.

The vagaries of the wind had spared British cities, and the Baltic convoy had escaped. Again, on the surface, John Paul Jones had failed in his designs. Yet the morale value at home and the prestige factor abroad of this second smashing victory over the vaunted Royal Navy were of inestimable value in the continuing struggle for independence.

British newspapers captured the nation's emotion at the height of the raid: "The appearance of Capt. Paul Jones on this coast has so increased the fears of the people of this city that they consider an invasion as inevitable. . . . They had 2000 sea and land forces with combustibles, prepared for setting fire ships or towns, but could not tell their destination . . . Not a day passed but we are receiving accounts of the depredations committed by Paul Jones and his squadron on our coast." Nor was the local military spared harassment, "There came an express to Sir John Irwin . . . that a fleet of French men of war was coming up the Shannon . . . the 32nd regiment and the 18th light dragoons marched immediately. However, in two or three hours more another express arrived, informing, that they were two of our men of war bringing their prizes in there." The recalled troops marched unhappily home.

As for the Royal Navy, the insufferable rebel gnat had now swatted three British men-of-war! His fleet was firmly blockaded in the Texel for the remainder of the conflict. It was another victory for Jones—without even fighting—as his battered squadron tied down a blockading force almost three times its size.

The war now entered a new phase, moving from *guerre du course* to *guerre d'esquadron*. French and British fleets and squadrons joined battle around the world, and the small Continental Navy became a mere sideshow of commerce raiders and dispatch carriers. John Paul Jones, with the navy that he helped birth, drifted into the shadowy corners of history. As he would die a pauper in Paris, so the ships of his navy would be scrapped one by one by a new and debt-ridden nation.

John Paul Jones was an exceptional commander. When he stepped onto the deck of his first command in the Continental Navy, he assumed the guise of raider by necessity. His small vessel, and its cohorts, could not stand against the more powerful ships of the British Navy; nevertheless, the strategy and tactics used by Jones to advance the American cause remain worthy of study.

Raids should be mission-specific. Jones operated with multiple objectives due to the need to capture, sink, and disrupt enemy commerce, but he never lost sight of his primary objective: to attack the enemy in their own safe harbors. This tied in directly to the underlying reasons for raiding: to reduce enemy military efficiency, to disrupt the enemy's economy, and (most important) to negatively impact enemy morale, creating a reduction in his will to continue the overall struggle.

Jones believed in raiding where it was least expected. Whether in the harbors of Nova Scotia or within the reserved British domain of the Narrow Seas, he realized that the safety of his vessel and the degree of impact depended on reaching those areas. To do so, he relied on subterfuge, and was not loathe to pass his ship off as a merchantman, nor to hoist a false flag.

Jones was a great commander, but he was not a great leader of men. Mutiny and near-mutiny seemed to follow his career from his earliest days as a merchant master. Yet Jones was not a cruel captain—"fond o' the cat," as the saying goes. His failure seems to have been an inability to identify with the needs and desires of his crew. Without that ability, he could not maintain the high unit morale recorded by the most successful raiders. If his men had fully supported him at Whitehaven, the resulting loss of shipping could have rivaled Cádiz, Pearl Harbor, or Taranto.

Jones also found his degree of success considerably diminished by the vagaries of nature. In his case it was fickle winds that died or blew from the wrong direction. With the age of sail long past, it is tempting to ignore mother nature. The raider does so at great risk. Suppose clouds had covered Pearl Harbor on December 7, 1941. Or simply consider the mud of the Eastern Front, the Khamsin of the desert, or a bird in a jet engine intake.

The mortal remains of John Paul Jones, raider by need and temperament, now rest at the Naval Academy in Annapolis, Maryland. He began a tradition that continues to this day. Throughout over 200 years of history, the men and women of the United States Navy have often been called on to assume the mantle of raider on sea, land, and in the air. They have never failed their nation and, like John Paul Jones, today's Navy could well say, "I wish to have no Connection with any Ship that does not sail fast, for I intend to go in harm's way. . . ."

4

The Cossacks at Hamburg

BY GEORGE F. NAFZIGER

Napoleon's Grande Armée had stumbled out of the snows of Russia in December 1812 and licked its wounds as it crossed the Grand Duchy of Warsaw and the eastern parts of Prussia. The Armée's peace was not long-lived, however, as the Russian army soon crawled out of the cold behind them, led by a pack of Cossacks that bounded after the struggling French, scooping up prisoners and loot like a pack of hounds chasing a beaten quarry to ground.

Napoleon's forces were in tatters, but the war was not yet over. His first order of business was to return to France to begin the process of rebuilding his army. When he departed in December, Napoleon turned command over to Joachim Murat, King of Naples, but the flamboyant cavalry leader had no heart for the job. Without Napoleon's permission, Murat took leave for Naples, turning command over to Napoleon's adoptive son, Eugène Beauharnais. Napoleon would not leave Paris to return to the Grande Armée until April 15. Eugène was on his own, with all that remained of the French army that had suffered so terribly in Russia.

Eugène struggled with his first major command. Napoleon bombarded him with advice and orders from hundreds of miles away, knowing little of the actual situation—and what he did know was a week old. Eugène was unsure of himself and responded to allied movements by withdrawing the French Army behind the Elbe, abandoning the Grand Duchy of Warsaw and Silesia to the Russians. Napoleon later subjected him to a terrible tongue lashing for this foolishness.

On the other hand, Eugène's withdrawal and timidity excited the Russian general Ludwig Wittgenstein, who was further intrigued by emissaries from Hannover, Oldenburg and Westphalia in Germany, who told him that revolts against the French were simmering in their cities. Wittgenstein saw a chance to wreak havoc behind the French lines and disrupt their communications. He dispatched three raiding columns, dominated by Cossacks, to the west.

Probably the most famous element of the Russian army, the Cossacks were a unique hybrid of Slavic blood, mainly Ukrainian and Russian peasant stock, and Mongol–Turkish culture.

From Roman times and before, Western civilization had been plagued by nomadic horsemen from the great plains of Eurasia, who would periodically burst forth to raid or, at times, to threaten the West with conquest. The word "kazak" was used as early as the tenth century to describe nomads of the Russian steppes like the Pechenegs and Kipchaks and other Turkic kinsmen of the Tartars. The word kazak is derived from a Turkish word meaning "freeman," though it also had the connotation of nomad or adventurer. Later the Circassians, another ethnic group from the Caucasus region, were known to both the Ottoman Turks and Russians as Cherkes; they were renowned as horsemean and raiders throughout eastern Europe. Before 1500, the word "cossack" came to be used for any steppe horseman, whether Tartar or Circassian.

From about 1450 onward, however, there took place a migratory movement of the Ukrainian population, generally farmers and fishermen, to the edge of the steppe, as they sought relief from Polish or Lithuanian oppression. Because these frontiersmen faced constant threats from Tartar raiders, they were forced to band together in military colonies for protection. Learning from their enemies, they soon developed light-cavalry tactics that proved effective against the Tartars, who, themselves, were considered to be devastating horsemen. The people of this new frontier culture were fiercely independent and would fight anyone who tried to control them. By the end of the 15th century, some of them began settling in the wild Dnieper River region and soon spread to other remote parts of southern Russia. It was among these rough immigrants that the modern Cossack culture was born.

As the principality of Muscovy began to expand south, it too faced the problem of how to combat Tartar raiders, who would ride around forts and infantry and against whom heavy cavalry proved useless. The Tsars struck a deal with the Cossacks, whereby these would serve as border guards between Russia and the wild eastern tribes of the steppes and the ever-threatening Turks. They also contributed men to the Russian Army whenever the Tsars called on them. For these services, the Cossacks were allowed to maintain nearly all of their cherished freedom. The agreement was mutually beneficial and lasted in one form or another until the Russian Revolution in 1917. (Lenin and Stalin could not tolerate the freedom-loving society which, in any case, fought mostly for the Whites during the Russian Civil War, and the Bolsheviks therefore virtually wiped out the Cossacks in a series of purges.)

At the time of the Napoleonic Wars, the largest single group of these fierce horsemen was the Don Cossacks. In 1802 the Don Cossack Army was established with the Guard Don Cossack Regiment, 80 line Cossack regiments (each of 500 men), the Ataman Don Cossack Regiment (1,000 men—which was raised only in wartime), and two horse batteries. Also in 1802, the Black Sea Cossack Army was organized with 10 mounted regiments and 10 foot regiments. In 1803 the Bug Cossack Army was created with three regiments, as well as other Cossack armies from Orenburg and the Urals. In August 1808, the Tchougouiev Cossacks became the Tchougouiev Uhlan Regiment, part of the regular army. A Siberian Cossack Army was formed that would consist at its height of 10 regiments and two horse batteries. In addition, a substantial number of other, smaller, Cossack formations were raised, particularly after Napoleon entered Russia in 1812.

The uniforms of the Cossacks varied widely and each of their armies was distinctive. The uniform of the Don Cossacks was the most "typical," if such a term may be used. Between May and September, the Don Cossacks wore a demi-caftan, or short vest, that stopped at the waist. In the winter they wore a full caftan or tcheckmen, a type of frock coat that reached to the knees. They wore long baggy breeches that, for the most part, extended over their boots. The entire uniform was always dark blue and had red piping on the cuffs and collar, and red stripes down the legs. The astrakhan, a fur hat with a red badge, was worn. It had neither a plume nor cords. The officers had no rank insignia prior to 1804 other than a white plume with a black-and-orange base. Their astrakhan had silver cords mixed with orange and black. In 1805 the Cossack officers were authorized to wear distinctive silver embroidery on their cuffs and collars.

The Cossacks' armament was not regulated, so they carried any number of pistols, swords, etc., but all carried a lance. The Cossack lance, which had a red shaft and bore no pennant, was the trademark of their profession and they plied it with deadly skill—much to the dismay of French stragglers.

In addition to the Cossacks there were other irregular cavalry forces in the Russian Army. Among these were the Bashkirs and Kalmucks, Mongolian tribesmen who inhabited the southwest of European Russia along the Volga. On April 7, 1811, the 1st and 2nd Kalmuck Regiments and the 1st and 2nd Bashkir were formed. No uniform was ever designated for these Asiatic tribesmen, and they were armed with bows and arrows. There is an instance of a French general at Leipzig scoffing at the horse archers as they approached him. When the laughing French general had his leg nailed to his horse's side by an arrow, he suddenly viewed them with a bit more respect. (All four of these regiments were disbanded in 1815.)

In 1813, Wittgenstein began plunging combined arms forces into the French rear to raise whatever havoc they could. The first such raids began in late March and would continue throughout the year. The four officers who executed the first raid—Czernichev, Benkendorf, Tettenborn and Dörnberg—were but the first of many such commanders. Their highly mobile forces seldom exceeded more than 2,000 men and, in addition to Cossacks, were comprised of elements of line cavalry (usually hussars or lancers), light infantry and some artillery.

They were raiders in the classic sense and not intended to stand and fight major formations. Instead they were to oblige Napoleon to spread his army around his rear areas and cover critical military positions and installations, as well as provide escorts for all important convoys of materiel. Though they were not strong enough to engage major formations, small groups of French reinforcements marching to the front could prove to be easy pickings. A battalion of 500 to 1,000 green infantry en route from their training depot to the Grand Armée could become quick prey for the veteran horsemen. At times, however, the raiders found themselves faced with much more substantial forces and were compelled to fight pitched battles.

Of the raiders' leaders, Tettenborn, Benkendorf and Dörnberg were all Germans who had defected to the Russians in the face of Napoleon's conquests, and been assigned to command Cossacks. Czernichev was the J.E.B. Stuart of the 1813 campaign, operating continually behind French lines and disrupting Napoleon's control over Westphalia and other regions of Germany. He later wrote an account of his victories, modestly referring to himself in the third person, although every time his name appeared it was highlighted in bold print.

In March 1813, Colonel Tettenborn's Cossacks were sent after the French in the direction of Magdeburg, and columns under Czernichev and Benkendorf moved towards Wittenberg in an effort to seize the French-held city of Hamburg.

Tettenborn's first stop was in Ludwigslust, the residence of the Duke of Mecklenburg-Schwerin. Here he found the duke ready to desert his alliance with the French and join the Russians. The duke formally joined the allied cause, quit Napoleon's satellite union of German states—known as the Confederation of the Rhine—and began to rebuild his armed forces (his previous, consisting of a single infantry regiment, had been lost fighting against the Russians a few months earlier). Now he declared that he would provide 2,000 infantry and 1,000 cavalry to support the allied cause.

Quitting Ludwigslust, Tettenborn marched on Hamburg, the major industrial and financial center of western Germany. Hamburg was unsettled. On February 24, the city had erupted in riots when a French customs official brutalized a citizen. Though unarmed, crowds attacked soldiers, customs officials and any other Frenchman they found. Peace was restored with bayonets and musketry, supported by some Danish cavalry. The situation, however, remained tense.

Hamburg's garrison, prior to the riots, had consisted solely of three infantry companies, some 300 men. After the riots, however, it grew with the addition of some gendarmes, a number of armed customs guards and two battalions of the 152nd Infantry Regiment, which were dispersed along the Elbe below the city. All told, this new garrison totaled about 800 men.

Not a single Frenchman was encountered as Tettenborn's Cossacks advanced to the Elbe, reassuring him that the French rear was largely unprotected. Word had not reached him, however, that General J. Morand was leading his division south from Swedish Pomerania towards Hamburg, while General Carra St-Cyr was en route to Bremen. Morand began his march in Stralsund on March 9, moving on Hamburg to quell the ferment and put down the rebellion.

When Morand was about 30 miles from Hamburg, the Danish general Ewald advised him that he had received orders from his king to oppose any effort by the French to occupy Hamburg in force. The motive of the King of Denmark in opposing his nominal allies stemmed from his fear that the war was being lost, and that he would need to reopen negotiations with the allies.

Ewald commanded 5,000–6,000 men on the frontier of Holstein, greatly outnumbering Morand's 2,500 to 3,000. Morand chose not to provoke the Danes and moved into Bergedorf, near Hamburg, on March 16, which placed him directly in the path of Tettenborn. Morand had barely established his bivouacs when the Cossacks descended on him. Unable to withstand their sudden attack, he quickly crossed the Elbe, abandoning his six guns to the Russians.

With General Morand pushed aside, Tettenborn had a clear march to his target, Hamburg, and he entered it triumphantly on March 18. Hamburg immediately reestablished its old form of government and proclaimed its independence from France (into which it had been incorporated in 1810). The flag of revolt was immediately raised in most of the lowland towns between the Elbe and the Weser. The newly independent Germans hustled the French officials who chanced to be there on their way, and cast every emblem and sign of French domination into the flames.

The jubilation was short-lived, however, and on March 20 cries of alarm sounded on the banks of the Ems. The cries soon subsided: it was the British arriving. The British, ever ones to meddle in Napoleon's affairs, had maintained an amphibious force on the island of Helgoland. They now landed a detachment in Cuxhaven and sent their ships up the Elbe to Hamburg. Yet another detachment landed on the right bank of the Weser near Bremerlehe, where they were joined by the coastal artillerists and the 7th National Guard Cohort, native Germans who had joined the uprising. The sight of red British uniforms on German soil also caused the gunners of Fort Blexen to defect.

The French response did not take long. General Carra St.-Cyr and General Morand moved quickly to suppress the revolt. St.-Cyr drew together the two battalions of the 152nd Infantry Regiment, placed Morand and his forces under his command and marched to stamp out the revolt. He dispatched two mobile columns, each of 1,200, men on March 23. One was sent against Bremerlehe and the other marched on Blexen.

On the 24th, the French column marching on Bremerlehe found itself facing over 1,500 ill-armed peasants and bourgeois under the command of an English lieutenant. After an hour of combat, the peasant rabble was crushed and all survivors executed without pity. The French then pillaged the village and, in a terrific rush, overran the traitorous coastal battery and their British accomplices. An officer and 19 soldiers were killed; a further 15 British soldiers were taken prisoner.

The second French column exploded into Blexen at noon. The peasants ran to arms, but could not hope to withstand the furor of the veteran French soldiers, and were quickly dispersed. The rebel gunners in Fort Blexen attempted to resist and fired a few rounds before they were overrun by a company of customs troops. The commander of the fort, a sergeant, was taken prisoner and executed. The French then pillaged Blexen and obliged it to pay a 15,000-franc war contribution.

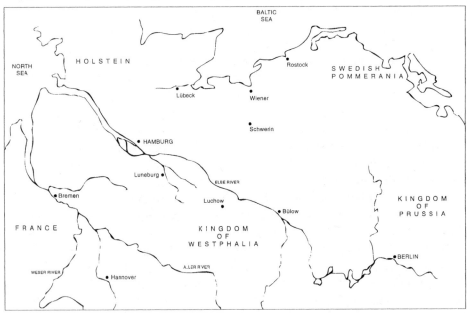

Areas where the Cossacks raided northern Germany, 1813.

But the French were not yet done, as an example had to be made to the rebels: the 20 prisoners taken during the action were marched into the cemetery and shot.

The second French column then marched into Bremen and made further examples, shooting the two leading citizens. Under the direct orders of Eugène and Napoleon, Carra St.-Cyr acted with swift and overwhelming violence to suppress any further thoughts of rebellion.

Meanwhile, Tettenborn, whose Cossacks had provoked this massacre, had quietly slipped his men across to the left bank of the Elbe, out of harm's way. Yet they continued to spread the revolt. The Cossacks marched past Bremen and danced about the flanks of St.-Cyr's forces. Tettenborn sent one of his officers to London to request that the British provide money, arms and munitions to support the creation of a corps of volunteers. Tettenborn's emissary proposed the formation of what would become known as the Hanseatic Legion, a force of 6,000 men of all arms. British help was also sought to raise what would become the Hamburg Civil Guard, which consisted of almost 8,000 men.

Russian General Wittgenstein had remained in Berlin during this period. When he learned of Tettenborn's entrance into Hamburg, he was delighted. He promptly organized a second force, which was drawn from the Prussians in Yorck's corps. This detachment consisted of three officers and 200 men and would be used to form the cadre of the Lübeck Battalion. A further force of 400 Mecklenburg Guardsmen moved into the city on March 28.

Wittgenstein now planned other mischief in Napoleon's rear and ordered Benkendorf to leave his position around Wittenberg to join Colonel Dörnberg, thus forming a force of 1,200 infantry and 2,000 cavalry. This force, placed under Dörnberg's command, was to move through Hannover and push as far as possible into the Kingdom of Westphalia, then ruled by Napoleon's brother Jérôme. Wittgenstein ordered Czernichev to support Dörnberg with his 2,000-man raiding force, which had been roving about in western Germany. Westphalian troops had deserted Napoleon in droves, and would continue to do so through 1813, so Wittgenstein justifiably believed the kingdom to be a ripe target where he could recreate his success at Hamburg. After Dörnberg united with Benkendorf, the combined force crossed the

Elbe on the night of March 25 and seized the small town of Werben.

Prince Eugène had meanwhile moved to Magdeburg, where he became involved in offensive preparations. He moved the three divisions of Lauriston's Corps (Maison—16th Division, Lagrange—18th Division and Rochambeau—19th Division) across the Elbe. Puthod's 17th Division remained on the left bank of the Elbe.

Marshal Davout was assigned responsibility for the region along the Elbe stretching from Magdeburg to Hamburg, including the 32nd Military Division, and Vandamme's corps was placed under his command. Napoleon felt that Davout knew the territory well and that his establishing his command in Hamburg would be "quite useful" to French war aims. On March 28, Davout arrived in Stendal. Sébastiani's cavalry corps and ten battalions of Davout's 1st Division were stationed along the Elbe near Stendal, to cover the French flank.

Prince Eugène was receiving more bad news than he could handle. Aside from insurrections in the north, and Cossack-dominated formations to his front and rear (which greatly unsettled him), he also learned of the Russo-Prussian alliance. The Prussian Army taking the field added still more pressure on the French, who were far from ready to deal with a fresh enemy.

Eugène feared that the allied armies were converging below Werben and that they would move on Hannover. He abandoned Napoleon's orders and moved Maison, Lagrange and Rochambeau's divisions to the Ohre, a river on the left of the Elbe a little below Magdeburg. He directed Grenier, Latour-Maubourg and the troops en route from Dresden there as well. Victor, commanding the 4th Division, was to remain on the lower Salle. Eugène had already forgotten Napoleon's lesson that the best way to protect Hannover was to threaten Berlin.

French reconnaissance forces were sent from Stendal towards Werben, and suddenly encountered Dörnberg's freelance raiders. General Montbrun, commanding 2,000 infantry and 500 horse, pursued Dörnberg as far as Neukirch, inflicting about 100 casualties on him, but doing little real damage. However, Dörnberg realized that he was greatly outnumbered by the French forces in the area. Supported by Czernichev, who had left the area before Magdeburg to the French, Dörnberg recrossed the Elbe on March 31, ten leagues south of Werben. He then occupied Dannenberg and Lückow. Czernichev occupied Wustrow.

Seeing Hannover open again, Dörnberg once more planned to "liberate" that city, but then he learned General Morand was approaching Lüneburg, just southeast of Hamburg. Lüneburg had been in revolt for about 12 days and had raised a few companies of poorly armed civil guards. Dörnberg and Czernichev felt that if they combined with the Lüneburg townspeople they could defeat Morand and so resolved to move to their defense. Unfortunately, when they arrived outside Lüneburg they found the French flag flying over the city.

General Morand's force had been reorganized in Bremen and now consisted of the Saxon Prinz Maximilian Infantry Regiment with two battalions (1,492 men), the 4/152nd Infantry Regiment (468 men), two companies of customs guards (147 foot and 25 horse), 3/8th and 17/8th Foot Artillery Companies (two 6-pound guns and 193 men), the Saxon battery of Hauptman Essenius (four 8-pounders, two howitzers and 170 men), and 45 gendarmes and cavaliers.

About noon on April 1, Morand had reached the barricaded gates of Lüneburg. His advance guard was greeted with a hail of musketry and he found himself faced by a small force of Cossacks, volunteer sharpshooters and a few Bürger militia companies in the terrain before the city. Morand sent Saxon skirmishers forward to drive them back and then reinforced these with two Saxon line companies and three cannon (2 French and 1 Saxon). The French and Saxons pushed forward rapidly, driving the Cossacks and burgers before them. Under the cover of this attack, Morand's column formed by pelotons at half-distance and began its advance. After the third cannon shot and, upon seeing the advancing French column, the Cossacks dispersed and the musketry diminished significantly.

Lüneburg was surrounded by an old wall pierced with five gates. Two Saxon companies under Major von Ehrenstein moved forward, cleared the barricades and seized the Neue Gate. From there they pushed into the city, chasing the Cossacks out the Bardowick Gate at the far end of the city. Simultaneously, the Saxon skirmishers, under Unterleutnant von Metzsch, attacked the old rampart situated between the Neue and Rouge Gates. They scrambled over the old walls and penetrated into the city. The 4/152nd Line Regiment moved to the north of the road while the remainder of the Saxon Prinz Maximilian Infantry Regiment, with Morand at its head, marched down the *pas de charge*—the main road—in dense column. By the time Morand's column reached Neue Gate, the rebels' fire from the ramparts had completely ceased.

Morand's column entered the city, drums rolling and flags flying. The Saxons established themselves in the city square, their regimental artillery and equipment in the center of their formation. The French installed themselves at the Sander Platz. The rearguard, which had remained on the heights to the west of the city, joined the main force after about an hour and a half, leaving their guns at the Neue Gate.

Sporadic sniper fire greeted the French as they entered the city, but they quickly and brutally stopped it. Dead civilians lay in the streets as Morand paraded through—townspeople who had foolishly opposed the veteran French and Saxon troops. The popular support within the city for the Cossacks caused Morand to take severe actions against the populace in order to impress on them that the French were in charge. A military commission was established and quickly ordered two burgers who were caught with guns in their hands shot. A further thirty of the most prominent citizens of the city were incarcerated.

Still, Morand knew that his small force was only an island of French power in an unsettled sea of enemies and rebels. Uncertainty about possible actions by the Cossacks and the burgers prompted him to establish defensive positions with an eye toward resisting a powerful attack. He did not establish any advanced posts because of the Russian cavalry superiority. Instead, Morand established garrisons at the city's gates, each consisting of an officer and 50 men. He also posted a single cannon at the Lüne, Bardowick, and Rouge Gates. The remainder of his forces—those who were not assigned to the gate garrisons—were deployed throughout the city.

At 6:00 A.M. on April 2, the alarm sounded when French pickets spotted Czernichev and Dörnberg's Cossacks approaching the city. They then remained at their posts until ordered to withdraw. Morand reacted to the nearby presence of Cossacks by establishing a reserve of an officer and 100 men in the town's citadel.

Dörnberg and Czernichev launched an attack at about 10:00 A.M., but it proved to be little more than a few random shots designed to harass and provoke the garrison. The French and Saxon infantry held their fire, waiting for a better target. At 11:00, as Major von Ehrenstein toured his posts, word came that strong columns of cavalry could be seen south of the city, advancing on the Sulz and Rouge Gates.

The allied column consisted of three Cossack regiments, a small unit of Bashkir horsemen, four squadrons of the Isoum Hussars, the 300-man Russian 2nd Jager Regiment, under Major von Essen, a 500-man Pomeranian Fusilier regiment, under Major Borcke, and a horse battery of Prussian guns. The Russian general launched a false attack against the southern and western gates (Rouge and Sulz), while he directed his main attacks against the Lüne and Oldenbrück Gates.

The column striking Lüne Gate contained von Borcke's Fusilier Battalion, three cannon and Benkendorf's cavalry. The force attacking the Oldenbrück Gate consisted of a battalion of the Russian 2nd Jager Regiment, three guns and Czernichev's Cossacks.

Around noon, Morand ordered Captain Pariset to take two guns, a small force of cavalry and some supporting skirmishers from the Saxon Prinz Maximilian Infantry Regiment to the heights located southeast of the town's tile works. He hoped that from this position they could take the Russian cavalry in the flank. However, as Pariset attempted to set up his battery they were themselves taken in flank by a four-gun Prussian battery standing to their east and charged by a squadron of the Russian Isoum Hussars. The Russians captured both guns and a major portion of the men, and the survivors were driven back to towards the Oldenbrück Gate. Soon afterward, the officers charged with defending the Rouge, Bardowick and Neue Gates sent word to Morand that they were under heavy assault. A reinforcement of 50 men was sent to the Rouge Gate and detachments of 100 men each were sent to the Neue and Bardowick Gates.

Saxon skirmishers lined the city walls and stood in its houses ready to engage the Russians. The 2/Prinz Maximilian Regiment was in the northwest part of the city, watching the Neue Gate and supported by two Saxon howitzers. The Oldenbrück Gate was protected by the 4/152nd Infantry Regiment and a single 4-pounder gun. Major von Ehrenstein, seeing the Rouge Gate under heavy Russian pressure, directed Lieutenant von Döring's reserve in the citadel to reinforce its defenders. At the Rouge Gate, the Russians were repulsed.

With the French and Saxons so far holding fast, Dörnberg split his forces into two columns aimed at, respectively, the Lüne and Oldenbrück gates. Six cannon placed 500–600 paces to the east of the town supported both attacks. The terrain allowed the Russians to advance quite close to the city without coming under significant fire. When they were near enough, they engaged the French garrison with musketry, supported by artillery fire, but their attack was not immediately successful. The French positions were covered by an arm of the Ilmenau River and it was necessary to force the bridges—which were under heavy French musket fire.

Prussian Major Borcke detached two platoons to face the Lüne Gate and moved the rest of his force (2 companies) against the village of Lüne. He intended to force his way across the first arm of the river, out of view of the French in the city.

General Morand was watching the battle from a position near the Lüne Gate and saw the attack of the allied infantry. He realized that the battle would require more than a few shots to chase off some overly aggressive cavalry, and called for the two Saxon reserve companies that remained in the city market. However, it was too late to stop Borcke. The Prussians successfully crossed the river and threw themselves at the gate's defenders. These were overwhelmed and nearly all taken prisoner. The breach was made and the allies began to pour into the city.

As the Prussians overran the gate, the two Saxon infantry companies from the reserve arrived. The Prussians exchanged fire with the Saxons, until they were ordered by Morand to withdraw. A terrific battle began within the city walls, now that the allies had penetrated the town. The allies were assisted by the city's residents, who fired on the French from their windows. The battle become one of hand-to-hand combat, a Prussian officer attempting at one point to seize the flag of the 2/Maximilian Regiment from the hands of Unterleutnant von Milkau. Though the Saxons retained their standard that time, it would not remain in their hands for long.

Leutnant Kunze arrived with two howitzers as the Franco-Saxon force withdrew into the marketplace. The guns were quickly unlimbered and began belching canister into the advancing Prussians. The Prinz Maximilian Regiment was now able to disengage and withdrew to the Neue Gate.

Morand realized that his situation was hopeless and rallied the troops at hand to effect an escape through the western gate. With 500–600 men and two guns he evacuated the town, leaving behind some troops still holding their positions and hundreds of others who had laid down their arms or fallen. The escaping French quickly formed a column and marched down the Harburg road, Morand hoping to reach a position on the heights to the west of the village, where he could establish a defensive position.

The French defenders at the Oldenbrück Gate had been more successful in holding off the allies, but did not receive the word to retreat. They continued to hold their position, fighting off the Russians until they found their position turned by the Prussians who had penetrated into the city behind them. Most of these men were obliged to surrender and only a few succeeded in rejoining their comrades.

The allied attack from the north, striking at the Bardowick Gate, consisted only of cavalry and these were never a serious threat. However, the Saxon garrison at the gate had been facing outward and did not know that the city's defenses had been broken. They quickly found themselves surrounded. Their efforts at a counterattack were in vain and they were forced to lay down their arms. Leutnant von Döring's forces, to the south, held out until the wounded Major von Ehrenstein was brought forward to tell von Döring that continued efforts were hopeless. Only the defenders of the Sulz Gate succeeded in escaping and joining Morand in his retreat.

General Morand's tiny force marched quickly, knowing that Cossacks lurked all around them, ready to pounce if the opportunity offered itself. Morand's column marched into Reppenstädt, a half league from Lüneburg. Part of his force consisted of 480 men from the Saxon Prinz Maximilian Regiment, supported by a 6-pounder cannon and a howitzer. The remainder was the 54th National Guard Cohort (also known as the 4/152nd Line Regiment). The entire force was surrounded by a screen of skirmishers.

Dörnberg had divined Morand's plan and had marched to intercept him. Once in Reppenstädt, Morand found himself under attack by Cossacks, the Isoum Hussar Regiment and four cannon. The initial charge by the Isoum Hussars was stopped short with heavy casualties in a hail of Saxon canister and musketry. The Cossacks, choosing not to act in mass, were held at bay by the French and Saxon skirmishers. The Russian artillery responded by firing canister on the escaping column and quickly dismounted both Saxon guns.

The allied advantage in cavalry convinced Morand that escape did not lie in marching across open country. He decided to return to

Lüneburg, where one of his other battalions still fought on. With Morand at their head, the Saxons charged forward, waving their flag, and crying "*Vive l'Empereur.*" They quickly overran the allied battery that had closed to fire on them, causing its gunners to flee and bayoneting those who were not fast enough to escape. But the Saxons, now reduced to 250 men by the detachment of two companies to face the Cossacks as they formed to attack, found themselves in a desperate situation. The 4/152nd had not followed their movement. Morand sent two aides de camp to bring them forward, but it was too late.

In the hail of fire Morand was suddenly knocked down by a shot. Major von Ehrenstein, who had already been wounded, assumed command and sent Captain Erdtel forward to Dörnberg to negotiate their surrender.

The 54th National Guard Cohort (4/152nd) and the two Saxon companies stood defiantly on the heights, surrounded by a sea of their enemies. Their muskets blazed at any who were foolish enough to venture too near, while they suffered under the allied artillery fire. Then their ammunition began to run out. Their position was gradually strangled by an ever-closing ring of allied cavalry.

In a last desperate gasp, the battalion commander cried "*En avant! Vive l'Empereur!*" The drums rolled, sounding the attack, bayonets were lowered and the battalion square lurched forward in one last effort to gain the shelter of the nearby village. But their morale was not up to the effort. Soon, cries of "Don't shoot!" rising from among the skirmishers were heard. They had run out of ammunition. The morale of the entire force collapsed before it had advanced 60 paces and the battalion surrendered.

Morand died surrounded by his men. The few hundred who remained with him surrendered immediately after he died. Many of the 400 Saxon prisoners deserted to the allies and enrolled in the newly formed Russo-German Legion. These men were taken to Königsberg, where the organization process began. The battle within and outside Lüneburg had lasted from noon to 5:00 P.M. A total of 3 flags, 2,300 prisoners and 7 guns were taken by the allies. The French and Saxons reported losing one officer killed and 11 wounded, but in fact about 130 men were killed and 220 wounded. The allies listed their casualties as 5 officers and 41 men dead and wounded.

On April 3, Dörnberg and Czernichev left Lüneburg and recrossed the Elbe. They had learned of the approach of French Marshal Davout's forces. Only a few hours after their evacuation from Lüneburg, the advance guard of Davout's corps—4,000 men of Lagrange's 18th Division under General Montbrun—marched into Lüneburg. The following day, Davout arrived at the head of 5,000 more. The lead elements of Vandamme's corps also began arriving in the area. Dufour's and Carra St-Cyr's divisions moved into Bremen and Dumonceau's 2nd Division occupied Minden.

Davout was charged with complete civil authority over the 32nd Military Division in order to grind out the revolt. He had the muscle to do the job. Vandamme, at the head of 28 battalions, and 33 other battalions were en route. A total of 61 battalions were dispatched to the 32nd Military Division to enforce French rule.

The raid was over and the Russo-Prussian corps withdrew to the east with little interference. Its job, to draw desperately needed French reinforcements away from the main area of conflict, had been successfully completed.

In response to this raid the French had sent a total of 32,109 men, including 2,219 sorely needed cavalry, into the 32nd Military Division. The 37th Westphalian Division (eight battalions, eight squadrons, sixteen cannons) and the Westphalian Guard and Kassel Garrison (2,200 infantry 1,100 cavalry and 1,200 gunners) were held in Westphalia to keep it under control. In addition, the 27th Polish Division, with 3,024 men, was also stationed in Westphalia.

If one considers the reallocation of men by the French and the losses they suffered during this raid, the actions of these six thousand partisans under their four Cossack leaders had the effect of shifting the effective manpower resources available in the main theater away from the French and to the allies by a total of 53,500 men. The total loss to the raiding parties was a few hundred casualties. The impact of this raid, though not well known, was probably decisive to the outcome of the 1813 spring campaign. In those coming months, even though Napoleon finally returned to command the army, he was not able to win a decisive victory, and in the fall Austria joined the allies. In October, at Leipzig, also called "The Battle of the Nations," Napoleon was defeated.

5

Mosby at Fairfax Court House

BY BRENT NOSWORTHY

Prior to the Civil War, a general once stormed, "This beardless youth ought to have been beaten over and over again; for who ever saw such tactics? The blockhead knows nothing of the rules of war. Today he is in our rear, tomorrow in our flank and the next day in our front. Such gross violations of the established principles of war are insufferable." The general was an Austrian and he was speaking of Napoleon Bonaparte.

This criticism of being iconoclastic and irreverent of existing military tradition certainly was never limited to Napoleon. Any military thinker or practician who tinkers with truly new methods is inevitably subjected to the same misunderstanding. John Singleton Mosby was no exception. Plying a form of the military trade that was barely tolerated and always looked down upon—despite continued success—the young partisan leader would encounter his share of detractors, and be criticized and ridiculed by friend and foe alike. Union soldiers derisively labeled Rebel partisans "bushwackers," "freebooters" and "guerrillas." Even Confederate cavalrymen, more often then not, disparagingly called their irregular confreres "Carpet Knights" or "Feather Beds," referring to the partisans' relatively easy lifestyle between guerrilla operations.

Despite a long series of successes, Mosby was subjected to similar attacks as late as early 1864. Complaining about the detrimental effects of partisans on regular troops as well as their general lack of effectiveness, in January of that year no less than General L. Thomas Rosser urged Robert E. Lee and the Confederate Secretary of War to disband all partisan organizations. He argued that all soldiers of the same rank be placed on the same "footing." Even those troops called upon to act occasionally as guerrillas would otherwise operate within friendly lines and conform, like everyone else, to the strictest discipline. In other words, everyone had to follow regular military regulations and the rules of warfare—even partisans and guerrillas. As Frank Williamson, one of Mosby's most loyal and competent followers, would later point out, ironically this would have soon accomplished "what the Federals, with all the resources at their command, after the most persistent efforts failed to accomplish—the destruction of Mosby's command." Fortunately, at least Lee and the Secretary of War held Mosby in much higher regard. Mosby and McNeill's bands of partisans were retained.

Mosby's partisans, like the others who would pop up in Virginia, Kentucky and Tennessee, had been legitimized by the Partisan Ranger Law that had been passed by the Confederate legislature. This awarded these roving bands the same rewards as had traditionally been extended to privateers on the sea. Though all captured mules and cattle had to be turned over to the Confederate government, the partisans were free to keep anything else they could capture. The most common items seized, of course, were Union horses.

Initially, Mosby's handful of men was comprised of troopers temporarily detached from his old regiment, the First Virginia Cavalry. However, as the success and profitability of Mosby's operations became known, at first a trickle, then a stream of volunteers offered their services to the partisan who was making such a name for himself. All types of men would offer their services. Those too young, or too old, for regular service were drawn by the promise of adventure and plunder. Often, regular soldiers on leave of absence would show up for a short time before returning to their proper regiments. Even those convalescing from wounds or illness would find the strength to join in for a raid or two. Mosby soon found himself leading such a motley collection of men and teenage boys that he affectionately referred to his command as his "conglomerates."

An increasing number of deserters were also drawn by the prospect of a relatively easy life and the possibility of near-instant reward. Mosby intuitively understood, however, that this not only represented completely unreliable material but would engender bad relations with the regular services—something that had to be avoided if he wished to curry good will from Stuart and his other supporters.

Between raids, to search for the enemy's weaknesses, Mosby's scouts and informants were continuously roving around the Fairfax/Centreville area, often riding as far as the Potomac River or the Shenandoah valley. As opportunities were uncovered, couriers were sent back to Mosby's headquarters. A rendezvous was established for his men to meet, and in a few short hours the band was ready for its next raid. The raid or expedition completed, the men scattered, most frequently among the farms along the Blue Ridge and the Bull Run Mountains. They had no fixed camp, since to settle down in one permanent position would have quickly led to capture.

While serving with the cavalry of the Army of Northern Virginia, from the very first Mosby was struck by what he perceived to be J.E.B. Stuart's new method of cavalry warfare. Instead of trying to mimic the traditional European cavalry predilection for massed confrontation of ordered formations on the battlefield, the mounted arm was to be used instead almost completely for unceasing reconnaissance to discover the location and strength of the enemy, to unravel his intended plans.

However, in later years after the war, Mosby would admit that it was Union general John Pope's proclamation prior to Second Bull Run that proved to be the real intellectual watershed that would direct and inspire his operations from 1863 onward. In his famous proclamation, Pope announced his intention to depart from traditional warfare, with its emphasis on "strong positions," "lines of retreat" and "bases of supplies." The Northern commander boasted henceforth that these defensive concepts would be completely unnecessary. He directed his officers to "study the probable lines of retreat of our opponents [Confederates] and leave our own to take care of themselves. Let us look before us and not behind."

Pope's intention, of course, had been to intimidate and humiliate his Southern opponents and at the same time to encourage his own troops. Ironically, it would provide a virtual blueprint of how to counteract his own operations. By disclosing his new freewheeling approach, the Union general had inadvertently revealed his "Achilles heel." Mosby immediately recognized the opportunity. If Pope and his staff didn't pay sufficient attention to their supplies and communications, he certainly would oblige them and perform that task himself.

At first glance, Mosby's plans seemed to be wildly optimistic, even bordering on delusional. After all, when he first started his partisan operations in early 1863 he literally had only a handful of men. Even after his famous raid on Fairfax Court House, when he was authorized to raise a battalion, the forces under his disposal were only a tiny fraction of the Union forces against which he would operate so long with such seeming impunity. But the young John Singleton came to a simple but profound observation about any military force that had to maintain lengthy lines of supply and communication. According to the young partisan leader: "As a line is only as strong as its weakest point, it was necessary for it [Union forces] to be stronger than I was at every point in order to resist my attacks." In other words, even though he would operate against an area containing many thousands of Union infantry and cavalry, he would always be able to find a point weaker than whatever forces he had available.

Mosby also had his own views about the various types of weaponry that were available to cavalry and the most appropriate tactics that should be used. After the war, Mosby would claim that he was the first cavalry officer to self-consciously discard the saber and rely exclusively on the six-shooter. His original regiment, the First Virginia Cavalry, were armed with sabers and Mosby "dragged" his around for the first year of the war. He cast it aside as soon as he had his own command. He and his men, when forced to engage the enemy, relied on a spirited charge, armed with two revolvers. According to Mosby, in his memoirs, his men "were as little impressed by a body of cavalry charging them with sabers as though they had been armed with cornstalks." In the same work he would claim that he had never heard of a saber being used to do anything other than to "hold a piece of meat over a fire for frying."

There is some evidence, however, that these views reflect more the attitudes of the writer and military authorities at the close of the nineteenth century than those of Mosby and his men in the early part of 1863. True, none of his men did carry a saber, but then few took along a carbine either. The simple truth was that sabers and carbines made too much noise. Remember, most of Mosby's operations occurred at night, under the blanket of darkness. The raiders whenever possible eschewed hard roads where the sound of the clop-clop of hooves could be heard for several hundred yards. To a successful partisan, the carbine was as useless as a saber, it made too much noise as it rattled against the saddle and was only good for long-range shooting. Partisans when forced to fight were almost always at extremely close quarters, where the quick-firing revolver was at a premium.

Mosby first conceived of his plan to wreak havoc against the Army of the Potomac while he was still serving with the First Virginia. He had performed a number of valuable services for J.E.B. Stuart and had direct access to this great cavalry leader. For months, he had hinted at, asked directly and then begged to be allowed to conduct partisan operations. Finally, Stuart gave him the go-ahead and temporarily assigned Mosby nine men from his regiment.

Mosby's first operation, in November 1862, was a terrific success and the ten men stampeded hundreds of Yankee soldiers near Manassas. Stuart was pleased and on January 24, 1863, assigned Mosby fifteen men and authorized his independent command. Over the next several weeks, Mosby and his tiny band successfully raided a number of weak outposts in the Fairfax Court House vicinity. The Union authorities were slow to realize these depredations were conducted by an organized band; they thought they were malicious, but spontaneous, actions on the part of local farmers sympathizing with the South. It soon became apparent, however, they were dealing with a clever foe, whom they then tried to catch. In this they were completely unsuccessful, and at least several Union officers were discredited as a result.

So far, Mosby had been operating with a handful of men. He was attracting attention and there was a slow trickle of recruits. Even so, the total theoretical force at his disposal was still less than fifty men.

All this would change one day in late February 1863. It all began quite inauspiciously when a deserter named James F. Ames from the Fifth New York Cavalry walked into a village suspected of abetting Mosby and his men. The large Yankee sergeant expressed an interest in joining the local Confederate partisans. He had been an ardent supporter of the Federal effort to maintain the Union, but now refused to fight for the abolition of slavery—which he believed was the new object of the war after Lincoln's Proclamation of Emancipation.

As luck would have it, Mosby and his men were there at the time when Ames walked up. Suspecting a Union ruse to infiltrate their organization, both Mosby and his men at first coldly turned down

Ames's offer. When he returned from a raid several hours later, E. Frankland, who would rise to a captaincy under Mosby, began to believe the Yankee's story. Taking pity on Ames, he brought the sergeant to an old widow's house where they messed together for several days. A large, muscular seafaring man with pleasing conversation and manners, Ames eventually impressed Mosby. Naturally enough, Mosby couldn't completely put aside his suspicions, but he was equally determined to put the stranger to the test.

Frankland had joined Mosby's band a few weeks earlier, in mid-February. Unfortunately, he lacked a horse and was unable to participate in any operations during this time. Mosby had promised to obtain a horse for him during his raids, but this never came to fruition, there being only enough captured horses to split among those who had taken part. Ames told Frankland he knew how to get horses for themselves. He and Frankland would simply walk into the Union cavalry stables at Germanton and take the horses they needed. Intimately familiar with the countryside, the location of all the Union cavalry vedettes and their daily routine, Ames was confident he could reach the stables undetected. Frankland agreed to the plan and the pair set out on their 30-mile trek on the night of February 28th.

They walked a good deal of the night, found a lodging and then continued their journey the next day. They finally reached their destination at 7 P.M. on the 29th, but had to wait until the Union camp settled down to sleep. They then casually walked into the camp, talked to a guard they encountered and then just as calmly saddled two horses and rode off, back to their neighborhood.

This simple but daring venture was to have a much greater effect on Mosby's group than the mere acquisition of two horses. The "Big Yankee" started to gain credibility among Mosby's men, while their leader immediately appreciated the potential value of Ames's knowledge. What Ames and Frankland had done to accomplish a limited goal, Mosby with thirty or forty men could do to achieve a truly audacious objective.

The commander of the Union cavalry in the Centreville-Fairfax Court House area was Acting Brigadier General Percy Wyndham, an Englishman who claimed to have served with Garibaldi. Wyndham seemed to take Mosby's depredations into his region rather personally and he had forwarded a number of threatening and insulting letters to the Rebel leader. The Confederate realized that by using Ames as a guide he could penetrate the Union patrols and vedettes, just as Frankland had, and under cover of darkness gain the center of Fairfax Court House undetected. Once there, he intended to "gobble up" Wyndham and Lt. O'Connor, a provost-marshal many of the Rebels and their sympathizers felt had gone too far in pursuit of his duty. If this were accomplished, it would prove especially embarrassing to the Union brigadier, and probably end his career, for he had already been captured the year before by Ashby in the Shenandoah.

Talking to Ames and drawing upon his already extensive knowledge of the area and Union operations, Mosby knew that there was a brigade-sized force of all three arms at Centreville. A line of outposts extended from there through Fryingpan to the Potomac. Another line of outposts extended along a line roughly running through Union Mills and Fairfax Station. There was a Union cavalry brigade encamped a mile east of Fairfax Court House and another strong cavalry outpost at Chantilly on the Little River turnpike. Mosby realized that there was a gap in the picket lines in the area between the Warrenton and Little River pikes and was determined to pass through the Union lines there. Ironically, Stoughton himself had become aware of this gap in the defenses and had communicated this fact to the officer in charge of the outposts. Unfortunately, nothing was done.

As Mosby would explain many years later, although his plan to walk into an area surrounded by thousands of the sleeping enemy, guarded by hundreds of pickets, seemed hopelessly unrealistic, its greatest virtue was its very audacity: to do what no one would expect or think possible. He also relied on the fact that no one would be able to distinguish between friend or foe on a dark, cloudy or moonless night.

On March 8, Mosby summoned his men to Dover in Loudoun County, and twenty-nine of his men responded. None was informed of his intentions and everyone assumed they were going out to attack a Union picket post—something which lately was becoming a habit. It was late afternoon by the time they crossed the Aldie gap and headed down the Little River turnpike directly towards Fairfax Court House. The weather was perfect for a raid. Snow was melting on the ground, which meant that the horses' hooves wouldn't make much noise. The drizzling rain obscured visibility.

As the raiding party headed down the road, Mosby turned to Ames, who was riding by his side, and quickly explained their destination. Mosby recalled: "Without being able to give any satisfactory reason for it, I felt an instinctive trust in his fidelity, which he never betrayed." They continued down the turnpike until they were within three miles of Chantilly. There they turned right and crossed through the gap in the picket lines. Silently they continued on, and finally reached the Warrenton Pike at a point midway between Centreville and their destination.

Mosby's plan had called for his small band of partisans to arrive at Fairfax Court House close to midnight. However, despite all his efforts to make certain that they stayed on their tight timetable, a seemingly minor wrinkle almost wrecked the entire scheme. Although it would hardly have appeared so to the small band at the time, the true crisis came unnoticed in the dark woods not long after Mosby and his men had left the Warrenton road. Darkness and the misty rain had already reduced visibility to a minimum. Matters only worsened once the men entered the dense forest, and the column frequently had to halt as those in front tried to find the way—sometimes literally groping about them. After one such short stop, the rear half of the column remained stationary, believing that Mosby had ordered a halt, and waited for further orders to advance. Unable to see more than a few feet in any direction, no one was aware that the two sections of the column had separated for fifteen or twenty minutes. Mosby and the front half of the band by this time had advanced a considerable distance. Still, there was nothing to do but turn back and try and find the missing men. Failure to do so would force them to abort the raid.

The men left behind were not offered such a clear-cut choice, and began to argue among themselves about what to do. A few wanted to immediately recross the Union picket lines and go back to Fauquier; others thought it best to remain were they were, convinced Mosby would send someone back. The majority believed that they should press forward and try to catch up to the commander, and this view won out after considerable discussion. Slowly feeling their way forward, fortunately they espied a faint glimmer among the pines in front of them a few minutes later. Doubling back, Mosby had stopped at a woodsman's hut, and this was the source of the light.

The entire force now together again, Mosby once more led his men towards their goal. The rest of the journey was without incident, the column only making occasional stops to cut whatever telegraphic wires they encountered as they neared Fairfax Court House.

The rebel band finally entered the town around 2 A.M. and silently made their way along the main street, which led to Little River turnpike in front of the village hotel. Mosby's men, of course, recognized where they were and knew they were in the center of hundreds, if not several thousand, of the enemy's troops. None balked, however. By now, they had confidence in Mosby's ability to lead them into, and then safely out of, the gravest dangers.

Everyone was asleep except for a few drowsy guards. Whenever they were challenged, Mosby or Ames would answer "Fifth New York Cavalry." It was so dark the sentinel couldn't tell the difference. Rebels would then approach the guard quickly and capture him so there would be no one behind to sound the alarm. Reaching Fairfax Court House, the thirty-nine men were divided into several squads. Some were sent to the stables to collect horses, others were to go to houses where officers were thought to be.

As a precaution, Mosby ordered that Ames and Frankland take a few men and proceed a little farther into the village. They had not gone far along the main street, however, when they were challenged by a sentinel pacing back and forth in front of the village's old hotel which, since the Union occupation, had been converted into a hospital.

Ames, having until recently served with the Union brigade, was totally familiar with its command structure and protocols. He casually replied that he was waiting for Major White of the Fifth New York. A plausible explanation, the Union sentinel was just about to return to his beat. Mosby, however, had given orders to capture and disarm all Union sentinels to prevent them from sounding the alarm. Ames bent over as if to whisper something in the soldier's ear, and as the sentinel returned and crouched over to listen, Ames and Frankland drew their revolvers, which they shoved in the surprised guard's face. As Mosby would point out in his memoirs, "a six-shooter has great persuasive powers" and the guard was made to agree that, at least in this case, discretion was truly the better part of valor.

As these events were unfolding, Mosby with most of his group quickly proceeded to the Murray House which they had understood to be Wyndham's headquarters. They arrived there only to find that the colonel's residency was in Judge Thomas' house, which they had passed as they first entered the town.

In Mosby's short absence, trooper Joe Nelson seized a telegraph operator and a soldier who had been sleeping in a nearby tent. Questioned, the soldier admitted he was a guard at General Stoughton's headquarters and proceeded to divulge its location. When Mosby rode up to his men beside the court house, he learned that the General was himself quartered in the town. Mosby immediately decided to change his plans. He would personally go after the larger "game," Stoughton, while Ames and a few trusted men would seek out Wyndham at Judge Thomas' house.

Ames and his party were able to make their way to the Thomas House and even into Wyndham's headquarters without being detected. Here, they were disappointed, however. That very evening Wyndham had boarded a train to return to Washington on official business. The bird they had been trying to catch had inadvertently escaped. Anxious to leave a message that they were able to enter Wyndham's inner sanctum, however, they snatched up his clothing and then gathered up the quality horses in the stables.

As the men searched the house, they found a few other occupants. Barging into another room, a Rebel private found a man asleep. Waking him, he extended Mosby's invitation for a Southern vacation. This man, who turned out to be an officer, was both quick witted and courageous, something probably no other Union officer in that town could boast, based on the night's events. He pleaded that he was only a civilian sutler and to substantiate his claim pointed to various confiscated goods that were lying about on the tables and chairs. He was getting away with his ruse, and the Rebel soldier was just turning to leave the room when Ames entered. As fortune would have it, the man in the bed was none other than the assistant Adjutant General, one Captain Barker, Company "L," Fifth New York, the very company Ames had served in until he deserted. Caught in his lie, Barker had to acknowledge his true identity and was made to quickly dress and accompany the Rebels.

As fate was handing Ames a happy reunion with his former superior officer, Mosby was making his way to Doctor Gunnel's residence.

A brick house, the Gunnell home was a little off the Little River turn-pike and stood on the western outskirts of the town. Describing the incident for a magazine article many years later, Mosby remembers he was accompanied by Joe Nelson, George Whitescarver, Welt Hatcher, Frank Williams and a man named Hunter. There was probably a sixth private, but his name unfortunately has been lost to posterity.

Like each of the other partisan squads stealthily working their way through the town, Mosby and the six others reached their objective unopposed and unnoticed. Reaching the house, Mosby and his men dismounted and approached the door. Boldly, Mosby pounded on the door, as if he was a courier seeking entrance with important informa-tion. A head soon appeared out of a second-story window and demanded who was there. "Fifth New York Cavalry with a dispatch for General Stoughton," was Mosby's instant reply. A few moments later he heard footsteps tramping down the staircase. The door opened and in front of them stood a man clothed only in his nightshirt and drawers.

Before the Union officer suspected anything was amiss, Mosby grabbed him by the shirt collar and whispered his true identity. As in many other cases, both on this night and in subsequent raids, this had the desired effect. The officer, a Lieutenant Prentiss, led his captors up the staircase to the general's room on the second floor. Mosby ordered Welt Hatcher and George Whitescarver to remain outside to guard the horses.

Quietly opening the door and stealing into the room, the five raiders lit a lamp. Given the enormity of the events that were unfold-ing in this room, and the complete lack of precedence, it is not sur-prising that every detail, every action would be permanently etched in the memories of all the participants. Even the dialogue, at least in sub-stance, has been preserved for posterity.

Lying on the bed before them was the catch: General Stoughton, sound asleep and seemingly dreaming. As Mosby looked around the room, it was immediately obvious to him that the general had been entertaining there. And, as Mosby would later observe, the uncorked champagne bottles lying about "furnished an explanation of the General's deep sleep." Stoughton had been hosting a number of Washington ladies who, as chance would have it, had left only minutes before Mosby and his men entered the town.

Drawing the covers off the sleeper, the Rebels expected the general to awake. The champagne was maintaining its effect, however, and Stoughton continued to snore loudly. Mosby had no time for ceremony, and possibly gave in to a touch of impishness. The general was rolled over on his side. Mosby lifted his nightshirt and delivered a single whack to Stoughton's buttocks. Mosby tells us the "effect was electric." Startled, the brigadier sat up and demanded to know what they were doing. Stoughton thought these men were couriers playing some sort of rude joke. Despite the dim light, he failed to perceive that these were Rebels, dressed in full Confederate uniform.

John Singleton leaned over the bed and calmly asked the general if he had heard of Mosby. Stoughton affirmed that he had, and asked eagerly if this Rebel, such a nuisance of late, had been caught. Mosby could not resist the irony of the question and replied, "No, I am Mosby—he has caught you."

Stoughton had a reputation as a gallant soldier and Mosby was afraid that he might try to escape, or at the very least attempt some sort of delaying tactics until the alarm was sounded and help arrived. Mosby told a lie, quite believable under the circumstances: the town was held by J.E.B. Stuart and Stonewall Jackson occupied Centreville.

Stoughton's face dropped as he realized the apparent hopelessness of the situation. There was nothing to do but acquiesce to his captors. He did ask, however, if Fitzhugh Lee, a former classmate at West Point was there and requested to be taken to his old friend. Mosby assented, but requested Stoughton, in turn, to dress quickly. Despite these injunctions, the general, who was widely known as a fop, dressed meticulously in front of the mirror, but in the excitement of the moment forgot his watch. Frank Williams picked up the forgotten item, which he handed to Stoughton as they exited the house.

Although Whitescarver and Welt Hatcher had been left to stand guard outside the Gunnel House and warn Mosby of any unexpected Union activity, they obviously could not resist the temptation of taking a more active role, like their comrades. Noticing some nearby tents, they snuck up and captured the seven occupants, who, they soon discovered, were troopers of the First Vermont Cavalry on duty as bodyguards to the commanding general. They also found two fine horses, bridled and saddled, and ready for any urgent dispatches that might need to be sent.

General Stoughton finally dressed and ready, the seven partisans and their prisoners quickly made their way back to the court house, the appointed place of rendezvous. Here, they found that the other Rebel squads had completed their assigned tasks, had returned and were mounted in preparation for a hasty departure. Mosby and his twenty-nine men had managed to round up about 40 Union prisoners and nearly 60 horses. Ames came up to Mosby and with great pride introduced Captain Barker to him. The former Union sergeant treated his former superior officer with the greatest civility and respect. Barker's initial attempt to avoid capture, as well as a desperate attempt to escape later, demonstrated that Ames's high opinion of this officer was not misplaced.

Miraculously, although the raiders remained in and about the town for slightly over an hour, not a shot was fired at them nor was the alarm sounded. At first, the raiders had been able to scoop up every Federal soldier they could find. Eventually, some Union soldiers spotted the Confederates and managed to escape the net. But time went by and there was still no alarm or organized attempt to fight back. Those who were not captured ran away to find a hiding place. Still, the situation was critical, and Mosby realized they would regain their nerve as soon as his band departed. Then the chase would be on.

It was now somewhere between 3:00 and 3:15 A.M., and the Rebel raiding party was finally ready to retire. There were actually more Union prisoners than there were rebels to guard them. The extreme darkness, however, once again worked to the raiders' advantage: the captured Union troopers were unable to distinguish friend from foe and had no idea how many Rebels there actually were. They thought the prisoners a little farther down the column were actually Confederate horsemen.

Wyndham having escaped capture, Stoughton became the desired prize and Mosby was determined to take every measure to prevent his escape. Stoughton wanted to mount one of the courier horses seized by Whitescarver and Hatcher. He was obliged to mount a horse of "lower mettle," instead. As an additional precaution, Hunter was ordered to hold Stoughton's bridle reins and under absolutely no circumstances to let go of these. Obviously, wisdom dictated that similar precautions be taken for all Union prisoners, but there were simply not enough captors.

After what probably seemed an interminable period of time, the raiding column finally was ordered underway. Quickly, but with as little noise as possible, it wended its way out of town. The clopping of the horses' hooves still could be heard aways, and as the column passed an outlying house an upper window was heard to be raised, followed by someone asking the name of the regiment in an authoritative tone. Emboldened by their complete success, the Confederate troopers laughed at the ludicrousness of the question.

The man in the second story turned out to be Lt.-Colonel Johnstone, commander of the 18th Pennsylvania Cavalry, who was spending the night with his wife. He realized that something was amiss and that the men in front of his house were almost certainly the enemy. Mosby realized, in his turn, that the authoritative tone of the challenge indicated the presence of a high-ranking officer, and, as such, another prize worthy of immediate capture.

Joe Nelson and Welt Hatcher, ordered to search the house, jumped off their horses, ran up and broke down the front door. Johnstone, meanwhile, had heard the column stop and had a premonition of what was to follow. With no time to dress, Johnstone ran down the stairs and fled through the back door clad only in his nightgown. Johnstone's wife, along with Captain Barker, proved to be the only ones on the Union side capable of "manly" actions that night. Just as Nelson and Hatcher were running into the house, Mrs. Johnstone managed to gain the hallway. Nonplussed, she stood her ground and, fighting "like a lioness," gave her husband the precious moments he needed to escape.

Her spouse, meanwhile, darted through the backyard, crossed some gardens and gained the stables and barn area. Though pursued by several of Mosby's rangers, he made his escape. The Rebels groped around in the dark, searching the hedges and barns for their elusive prey. Others ran into the house and searched for spoil. Coming to Johnstone's bedroom, they spotted his uniform, hat and watch, which they grabbed as consolation for not catching the fugitive. Nelson quickly claimed the hat to replace the one he had lost on another raid a few days earlier.

This little episode had cost the raiders another fifteen minutes or so, pushing them even further behind Mosby's original schedule. Time was fast running out if they hoped to leave the town, pass by the cavalry camps and slip past the vedettes before the break of day. Mosby ordered the search called off, the men remounted and the column was set in motion once again.

In the various reports that had to be written in the days following Mosby's raid, there were some subtle and some not so subtle attempts to paint a different picture of what had actually occurred and how each major Union participant behaved. The Marshal-Provost's report, for example, would say that Johnstone hid under a barn. Evidence suggests that he actually climbed under an outhouse, a place unlikely to be thoroughly searched. In any case, the panic-stricken Johnstone remained in his hiding place the several hours until dawn, not realizing that his pursuers had long since left. To add injury to insult, when he finally gathered up the courage to emerge, a nail ripped off his nightgown as he slid under the boards.

What Johnstone did next cannot be ascertained with definite certitude. Some say he ran to headquarters in his nude and noticeably odoriferous condition; others write that he ran back to his wife. The latter, although certainly glad to see her husband uncaptured and unhurt, politely declined embracing him until he was washed. Only after he been scrubbed with a horse brush and curry comb was he finally ready to pursue the long-gone Rebels. Then he led his men in the opposite direction.

Mosby, meanwhile, after departing Fairfax Court House had put into motion his plan to confound the pursuers who would inevitably follow. For about half a mile his column headed towards Fairfax Station. At this point, the column wheeled 90 degrees and made for the road that connected Fairfax Court House to Centreville. This was probably the last direction the Union pursuers would expect him to take since it led to camps containing several thousand Union cavalrymen. Mosby's plan was to come close enough to the cavalry camps to confuse anyone who attempted to follow him from Fairfax Court House, but remain distant enough so as not to disturb the sleeping soldiers and their unvigilant vedettes. Mosby's men had carefully cut all the telegraph wires and no one outside the victimized town had any inkling of what had taken place. Of course, for Mosby's plan to work it was absolutely necessary that his party wind past Centreville and its outlying fortifications while it was still dark.

As they rode along, Stoughton at first was able to uphold an even disposition; he expected at any moment to hear the sounds of his own cavalry, who would capture the heavily laden Confederate raiders. Talking to Mosby, he even acknowledged that the Rebel captain had performed a truly brave exploit. A few moments later, they reached the pike about halfway between Fairfax Court House and Centreville. Mosby would later recall feeling that they were now relatively safe from pursuit. The danger now was posed by the camps and the guards in their front.

Up to now, the column was a relatively loose formation. A few of Mosby's men had ridden as flankers; others were slightly to the rear. It was extremely dark and most of the prisoners were holding their own reins. The men at the column's sides and rear were positioned to prevent escapes, but they were only partially successful and a number of Union cavalrymen, including Lieutenant Prentiss, Stoughton's aide, managed to get away. Reaching the Centreville pike, however, the column was now closed up and Mosby rode a distance on the pike to reconnoiter. The way clear, Mosby ordered Hunter to bring the column forward at a trot. Now only four miles from Centreville, the column continued forward. Nelson and Mosby frequently would stop and listen for the sounds of pursuit; all they could hear was the hooting of owls.

Dawn was starting to break. The critical moment had come. They still had to pass by Centreville, only a mile away, and the outlying picket posts. Upon galloping back up to the column, after one of his halts to listen, Mosby found that it had come to a stop. Ahead about a hundred yards could be seen a smoldering campfire beside the road, where a Union picket had spent the night. Hunter cautiously went up to investigate. The post was now deserted. As Mosby had counted on, the officer had ordered the guards to return to camp at dawn.

The column resumed its march along the road, in full view of Union fortifications 200 or 300 yards away. One of the sentinels on a parapet saw the column and called for it to halt. Mosby had no intention to comply; however, just at this moment his attention was diverted as a gunshot rang out. Looking ahead, he could see that Captain Barker had put his spurs into his horse and was galloping hell for leather towards the fort. Before he could reach safety, his horse stumbled into a ditch and he was thrown onto the ground. Mosby and some of his men quickly rode up to Barker and helped him remount his horse—unhurt but still a captive. Meanwhile, the Union sentinels remained passive, despite the sound of a gunshot and despite the failure of the column to halt as ordered. It was still very gray and hard to distinguish the colors of the uniforms. The column was riding away from the cavalry camps and the guards assumed it was a Union cavalry patrol heading out on one of the frequent early-morning patrols.

A few minutes more, and the raiding party was finally past Centreville. The last hurdle was to cross Cub Run, after which they would be outside the reach of any pursuit. As they neared the stream, they saw that their difficulties were not over. Melting snow had swelled the stream and it was now "booming"—overflowing—in a raging torrent. The choice was either to turn back and find some other way back to the lines or to try to wade across. Looking over his shoulder, Mosby could still see the cannons "bristling through the embrasures." He plunged in. His horse held its own and he made it across. Stoughton was next. Then the rest of the raiders and prisoners plunged into the stream. All made it across, although many were driven slightly downstream, and all were soaked and wet.

Once over, they were now relatively safe. Circling slightly back, they regained the Warrenton Pike. Soon they passed the very spot where Stonewall Jackson had repulsed Fitz John Porter the previous year. Climbing a high hill, they were able to see miles back. They were not being pursued. Up to that point it had been a dreary day, and although last night's drizzling had long stopped, it remained overcast. As the column stopped atop the hill, however, there was a brilliant sunburst. Mosby could not help but be moved by the moment. His career was on the ascendant; this was his Austerlitz, and in his own words, "He had won the lottery of life."

Frankland had been sent ahead to Warrenton to get provisions. The men had not eaten for many hours and they were both extremely tired and hungry. As the main body of the party approached, the townspeople came out with a breakfast they had prepared. The men, women and children cheered heartily. This was Mosby triumphant. Mosby and some of his men, with General Stoughton, stopped a while at the Beckham House. Stoughton had gone to West Point with one of the Beckham sons and had vacationed there; the family gave him a kindly welcome.

The next morning, Mosby and his men arrived at Fitzhugh Lee's headquarters at Culpeper Court House, bringing with them one general, two captains, thirty privates and fifty-eight horses. Mosby would never forget the surprised expression on Fitz Lee's face as he introduced Stoughton to his old classmate. Lee said little to Mosby, who sensed his coldness. Truly his star was rapidly rising; it was not surprising that it would awaken jealousies. Discreetly, Mosby left the room without further adieu and went down to the train station to greet Stuart, who had just arrived from a court-martial at Fredericksburg. Unlike Fitz Lee, Stuart was openly delighted with Mosby's raid, and over the next several days did everything in his power to ensure that the Confederate authorities knew about the daring accomplishment.

The great raid was over. John Singleton Mosby would soon become a household word throughout the Confederacy. Largely as a result of the raid, he was promoted to major and authorized to raise a battalion. Finally, his dream of engaging in partisan war on a grand scale was materializing.

As would be expected, the consequences of that March 8th night on the major Union players were quite the opposite. Although he would soon be exchanged, Stoughton never returned to the service, for he could never live down the ridicule stemming from the manner in which he had been captured. Hearing about the fiasco at Fairfax Court House—including the capture of Union horses, soldiers, officers and a general—Abraham Lincoln commented, "Well I'm sorry for that. I can make new brigadier generals but I can't make horses." Colonel Johnstone was also disgraced. He could not live down the ignominy of showing up at headquarters in the nude, as well as his choice of hiding places. Although Wyndham was not present during the raid, his inability to control Mosby became known and he too was relieved of command.

In the days following the raid, Union authorities would investigate what happened and the reasons for Mosby's remarkable success. They erroneously concluded that it was an "inside job," believing that Southern sympathizers in the town had passed on critical information to the partisan captain. A few suspects, particularly Antonia Ford, the young woman who had lodged Stoughton's female guests that night, were arrested and sent to a prison in Washington D.C. There was no basis for these charges, however, and they would eventually all be released. Probably the most lasting result of Mosby's daring raid subsequently took place in the marital, rather than martial, arena when Antonia Ford married a Yankee provost marshal involved in the case named Joseph Willard. She "got her revenge," Mosby was heard to remark.

6

Morgan Across the Ohio

BY STEVEN M. SMITH

In July 1863, a quiet night in southern Indiana was suddenly broken by the clamor of church bells—not the cadenced peal that announced the beginning of services, but a more urgent, non-stop summons. Men and women hurried to dress and rushed expectantly into town. The news that first week of July had so far been nothing but good—huge Union victories in both the east and the west—and in more than one Hoosier's mind, as they ran to join the crowd, the bells might possibly herald the best news of all: the South had finally fallen. But on the night of July 8, the people of Indiana would find their jubilation turn to panic, their complacency to alarm. Their peaceful countryside, in fact, had become a potential battlefield: Morgan had crossed the Ohio.

After two years of the Civil War, the name John Hunt Morgan—along with that of another Confederate commander, Nathan Bedford Forrest—inspired fury and dread among partisans of the Union, and wild admiration among the people of the South. Largely operating apart from the main armies, the two cavalry leaders were providing an answer to the Union occupation of Kentucky and Tennessee: surprise, devastating attacks behind the Yankee lines.

Morgan himself was a handsome, 38-year-old six-footer, who cut a sartorially impressive figure and displayed a gay, courtly manner. A veteran of the Mexican War, he had captained a pre-war company of militia that was quickly converted into the 2nd Kentucky Cavalry after the start of the War Between the States. At Shiloh, Morgan's men had fought as traditional cavalry, supporting the infantry with attacks of their own in that maelstrom of blood and confusion. But then, during the lull that followed, Morgan received permission to act independently, staging raids against Union lines of supply. And in these endeavors he and his men were rewarded with spectacular success.

Although, as seen through Northern eyes, Morgan and Forrest were essentially a two-headed monster, their characters were significantly different. Forrest, a Tennesseean, defended his homeland more ruthlessly than romantically, and with a fine-honed sense of strategy against Yankee trepidators. The ex-slave trader ended his career having slain thirty-two men personally, his command accused of atrocities and Jefferson Davis ruing the fact that the Confederates had not earlier recognized his military brilliance.

Morgan, by contrast, resembled not so much a medieval Celtic avenger as a knight gallant, who fought for hearth, womenfolk and bluegrass, including the God-given right of the people of Kentucky not to be overcome by invaders. Ironically, since Kentucky was a "border state," Morgan's men would often find as their opponents other Kentuckians, dressed in blue, fighting for exactly the same thing.

Morgan's first independent foray, into Kentucky in 1862, began promisingly when his command of just over 300 men attacked a 400-man garrison at the town of Lawrenceburg, taking 268 prisoners. He then attacked the 600-man 1st (Union) Kentucky Cavalry at Lebanon. After a hard fight, Morgan's men had wounded and captured the 1st Kentucky's commander, Colonel Frank Wolford. Morgan rode over to "Meat Axe" Wolford, whom he knew and respected from the Mexican War, to inquire about his wound and to offer a parole. Wolford declined the parole and was then rewarded for his stubbornness when the 7th Pennsylvania Cavalry came charging to his rescue, forcing Morgan's men to scatter.

After his retreat, called "the Lebanon races," Morgan gathered up the remnants of his command and rode to Caveburg, Kentucky, where he destroyed a Union freight train. Another train, filled with troops and their wives, approached and the Rebels stopped it, forcing its occupants to surrender. One of the women pleaded with Morgan for the safety of her officer husband, to which entreaty the Confederate acceded in typical (if not apocryphal) fashion. "My dear madam," he said. "He is not my prisoner. He is yours."

When Morgan returned to the Confederate lines after his first Kentucky raid, he was already celebrated in the press and had no difficulty recruiting additional men for his cavalry. Morgan's command already included 25-year-old Basil Duke, an exceptionally clear-headed officer and his right-hand man. It also included the scout Tom

Quirk, the (Kentucky-born) Texan Richard Gano and the ex–New York editor Gordon E. Niles, who produced the first issues of Morgan's newspaper, *The Vidette* (published "semi-occasionally"). The saturnine telegrapher George Ellsworth, who had studied at Samuel Morse's school, could perform tricks on the wires with his portable break-in tools that rivaled the touch of a piano virtuoso. Perhaps the most valuable addition to Morgan's men in the summer of 1862, however, was George St. Leger Grenfell, who bore letters of introduction from Pierre T. Beauregard and Robert E. Lee.

Grenfell, then nearly 60 years old, was a Britisher who had fought as a teenager with the Moors in Algeria against the French. Then, having no particular axe to grind, he had joined the French against the Moors. He had fought for the Turks, with the British in India and also for his country in the Crimea. After service with Garibaldi in Italy, he had settled in South America to raise sheep when he heard about the new war in America. He thus sailed into Charleston harbor, mentioning to officials in Richmond only that he preferred the cavalry branch.

Although the Kentucky boys in Morgan's command already knew how to ride and shoot, St. Leger Grenfell was able to teach them new techniques about how to handle horses in the midst of gunfire, as well as maneuvering as a unit and standard drill. Morgan was not a disciplinarian—Duke was better—but "Old St. Lege" was the man who transformed Morgan's raiders into an elite command.

Like Forrest, Morgan could coordinate with the army, performing the kind of duties that had been assigned to cavalry for hundreds of years. But on independent operations, the raiders fought instead as mounted infantry, employing their horses for mobility and speed, but engaging in combat on foot. Their weapon of choice was a sawed-off Enfield carbine, as well as Adams' or Colt revolvers (often a brace). Every man owned a knife and some, including Morgan occasionally, carried a sword.

On their second raid into Kentucky, in July 1862, Morgan's men heard en route that Union officer Thomas Jordan of the 9th Pennsylvania Cavalry had been abusing the women of Sparta, Tennessee. Jordan had threatened the women that, unless they cooked for his command, "they had better sew up the bottoms of their petticoats." The next morning, the 9th Pennsylvania was surprised by a three-pronged attack by Morgan's men that captured their camp, most of their guns and horses, plus 300 soldiers, including the offending

Major Jordan. By the time the July raid ended, Morgan's 900-strong command had destroyed Federal armaments and supplies in seventeen towns, scattered 1,500 militia, captured 1,200 Union regulars and hundreds of horses—all at a cost of about 90 of their own in killed, wounded and missing.

By August, the entire Confederacy was on the counteroffensive, moving into the border states. Lee was marching into Maryland and two Rebel armies, under Kirby Smith and Braxton Bragg, were advancing in Kentucky. In support of Bragg, Morgan was launched against Union communications above Nashville, taking the town of Gallatin, capturing 300 men and then destroying twin railroad tunnels to close a key Federal supply route for months.

While operating in Tennessee, Morgan received word that Union Brigadier General Richard W. Johnson had vowed to "bring him back in a bandbox." Don Carlos Buell had given Johnson the pick of the Army of the Cumberland's cavalry with which to find Morgan, and these—elements of the 2nd Indiana, 4th and 5th Kentucky, and 7th Pennsylvania—were encamped at Hartsville. Morgan attacked the town and dispersed the Union cavalry on August 21, capturing General Johnson. The Confederate high command, thinking Morgan might have been in trouble, ordered Nathan Forrest to ride to the scene. When Forrest arrived, a day later, he regretted having missed the battle, but he and Morgan celebrated nevertheless, enjoying a pleasant evening of conversation.

On September 4, Kirby Smith entered Lexington, Kentucky, Morgan arriving a day later in his hometown to great acclaim. It was at this time that Morgan's men first began to contemplate the idea of a raid into the North. (Basil Duke even tried, with a limited force, but was fought back at the banks of the Ohio, at Augusta.) After Fort Donelson, Shiloh, Corinth, New Orleans and other defeats in the west, the South had now bounced back, all the way through Kentucky. To Morgan and his lieutenants, it was already time for the North to feel the war on their own ground.

Among Southern sympathizers in Kentucky and Tennessee, the actions of John Morgan had already begun to transform themselves into legend. To people forced to live under the Union heel, the sight of Morgan's men riding freely and fearlessly among the occupying forces, defeating them in battle and destroying their stores and supply routes was exhilarating. Even for those who never saw the raiders, news of their exploits, carried regularly in Southern newspapers, lifted morale. Outside Castleton, Kentucky, a teen-aged girl named Sally McCann witnessed a detachment of Morgan's men camping one evening on her father's farm. She later wrote, "I felt that night as if I were living in the time of Robin Hood."

The Northerners' opinion of Morgan was entirely different. "Cowardly, murderous, bushwhacking, horse-thieving plunderer" would have been closer to their view. Of course, this opinion resulted largely from frustration.

The "high tide" of the Confederacy, as the early fall of 1862 has been called, was short-lived. Maryland had not risen en masse to the Rebel cause and neither had Kentucky. Bragg fought an inconsequential battle at Perryville and then determined to withdraw back south. Lexington was once more abandoned.

Morgan nevertheless continued to operate as far north as he could. He returned to Lexington and captured its garrison, the 4th Ohio Cavalry, in October. After taking several Federal wagon trains and tearing up large tracts of railroad, he came back to the Confederate lines; but then, in November, he was at Nashville, attempting to set fire to a Union rail marshalling yard. Now commanding almost 4,000 men, on December 9 he took Hartsville, which had been held by 2,000 Federals, capturing 1,800 and killing or wounding the rest. At Elizabethtown, he captured 600 Federals, and then 650 more, holed up in two stockades, when he destroyed the train trestles over Bacon Creek, an important supply route which ran through a ravine above Nashville.

On New Year's Eve, Morgan's men captured a few Union wagons and were able to enjoy a satisfying repast in Tennessee. However, that same day, many thought they heard the rumble of cannon fire from over the hills. Braxton Bragg was, in fact, fighting the battle of Murfreesboro—without his best cavalry, neither Morgan nor Forrest. The battle, which resumed and concluded on January 2, was a tactical victory for the South, but Bragg's subsequent retreat, as after Perryville, rendered the outcome moot—underlining, too, the pointless waste of 13,000 lives. The much-criticized Bragg was a brave, steady general who ultimately failed his cause only because he lacked the brilliance the South desperately needed in its commanders, to offset other Union superiority. To Bragg, however, the absence of thousands of his cavalry at Murfreesboro was a source of deep resentment, and he was never to view the endeavors of his raiders so benevolently again.

Morgan spent the next few months screening the army, engaging in repeated skirmishes with aggressive Union cavalry, and his command was partly dispersed as far-flung pickets and scouts. Meanwhile, the coming summer promised to be the climactic season of the war.

By mid-1863, the Confederacy was in deep trouble in the Mississippi theater of operations, where Union General Grant was methodically destroying every last vestige of Rebel resistance. In the east, on the other hand, Robert E. Lee continued to conquer, and was unconquerable, against the largest Federal armies. But in the western theater, including Kentucky and Tennessee, the situation had become stagnant, marked only by a continued decline in Southern resources. In June, Morgan's spirits were lifted when Bragg finally sanctioned another raid deep into Kentucky. Joseph E. Johnston had been dispatched from Bragg's army to attempt to lift the siege of Vicksburg, thus necessitating a strategic pullback in Tennessee. To distract the Federals and prevent pursuit, Morgan was requested to penetrate the Federal rear to create what havoc he could. To Morgan's request that he be allowed to cross the Ohio River and raid the North itself, Bragg firmly refused. Morgan was to raid Kentucky and then return to the army.

Morgan, however, for a variety of reasons, had every intention of disobeying his orders. The obvious reason was that, with Union armies occupying (depridating) whole sections of Southern states, the compulsion to somehow hit back at Northern territory was strong. (This feeling may have been accentuated among Morgan's men because most of them were from Kentucky, a state with Union neighbors.) Further heightening Morgan's ambition was news of Union General Benjamin Grierson's raid in April. Grierson, with 1,700 cavalry, had ridden from southern Tennessee through Mississippi, emerging at Baton Rouge, Louisiana to announce, "The Confederacy is an empty shell." The news of Grierson's raid did not sit well with Morgan's men, who were probably the only Rebel unit capable—and anxious—to retaliate in kind.

The final reason for Morgan to penetrate the North—and one that provided an operational rationale—was that Robert E. Lee and the Army of Northern Virginia were just then invading Pennsylvania. If Morgan's men could cross Indiana and Ohio, and join hands with the victorious Virginians on Northern soil, that could, possibly, become the crowning moment of the Southern cause in the war.

The raid began on July 2, 1863, when Morgan's 2,460 men forded the Cumberland River at Burkesville, in southern Kentucky. The command consisted of ten regiments divided into two brigades, under Basil Duke and Adam Johnson, plus artillery of four Parrot guns and two mountain howitzers uner the command of Captain Ed Byrnes. After dispersing pickets along the river, the Rebels were attacked by strong Union forces, though only part of Federal strength in the area. Morgan easily broke through and headed due north.

On July 3, three regiments of Ohio cavalry, led by Frank Wolford, attempted to confront the invaders, but were surprised by the Confederates' strength. "We thought it might be a small force we could crush," said an Ohioan. "But when we fired musketry we were answered with grape and cannister; when we fired a few rifle shots we were answered with whole volleys of musketry. We speedily beat a hasty retreat, going as fast as our horses would carry us."

On July 4, the Rebels came upon the 25th Michigan Infantry, strongly entrenched behind a log barricade at an elbow in the Green River called Tebb's Bend. Called upon to surrender, the Michigan officer replied that on any other day he might, but "The Fourth of July is a bad day for surrenders, and I must therefore decline." The Confederates suffered 36 killed assaulting the stronghold before Morgan called off the attack. The Rebel raiders were more interested in the Ohio than in a log fort in Kentucky.

Continuing north, Morgan took the town of Lebanon, capturing 500 men, and then Bardstown, where the raiders captured or dispersed some companies of militia. By now they were almost at Louisville, on the Ohio River, and, according to Morgan's telegrapher, George "Lightning" Ellsworth, the Federals were in an uproar.

Of all the colorful characters in Morgan's command, few were as vital to his success as Ellsworth, the scrawny telegrapher who, with his portable battery and instruments, could eavesdrop on reports of Union troop movements, while using his own genius for imitating other operators' styles to issue false commands. Typically, Morgan's men would cut a swathe through enemy territory fully informed of the positions of Union units ahead, while ripping up all the wires behind them. Ellsworth would meantime "dispatch" Union cavalry to any location he or Morgan deemed advisable, as long as it was in the wrong direction.

On one occasion, Ellsworth had a close brush with death when he "borrowed" St. Leger Grenfell's horse to chase a bushwhacker who had gotten under his skin. Ellsworth was outfoxed by the partisan, however, and in the process lost the horse and its saddle, which contained all of the Englishman's money. Ellsworth had to hide for three days while "Old St. Lege" went looking for him, sword in hand, the mercenary's face the same color as his red fez.

As Morgan's men raced north, the city of Louisville braced for invasion. Some Union formations, thanks to Ellsworth, rushed to defend the state capital of Frankfort. By this time, Ellsworth was well known to the Federal high command, as evidenced by a plaintive report on three Federal telegraph lines made by Major General Gordon Granger to William C. Rosecrans: "Two of the Louisville lines are out and Ellsworth is on the other."

Morgan was, in fact, turning due west, towards Brandenburg on the Ohio River. On July 7 he had sent ahead companies under Sam Taylor (a nephew of Zachary Taylor) and Clay Merriweather to seize boats for the crossing. Finding an old wharfboat at the docks, forty raiders hid themselves behind its gunwales and then floated to midstream, where they captured a passing steamboat, the *John B. McCombs*. Then, having boarded the steamer, they struck again into midstream to intercept a mail packet boat, the *Alice Dean*.

When Morgan's division rode into Brandenburg, their transportation was moored and waiting. Captain Tom Hines, who had been sent to reconnoiter southern Indiana, meanwhile reported in at Brandenburg with his scouts. At this time, Morgan informed his entire command that they were going to raid the North. Cheers erupted down the length of the entire column.

On July 8, the morning fog lifted to reveal about 100 militia on the Indiana side, surrounding an old parade-ground cannon. The cannon actually wounded an assistant quartermaster in Johnson's brigade; however, the musket fire of the defenders failed to reach across the half-mile-wide river. Morgan's Parrott guns sent over a few shells and the Indianans hastily withdrew.

The *John D. McCombs* and the *Alice Dean* had deposited the first Confederates on the Indiana side when a Union gunboat suddenly came around the bend. The Union vessel began firing, first at the northern side of the river, then at the southern, but Morgan redirected his Parrotts to fend off the intruder. After about an hour, the gunboat broke off the battle and retired downriver. It was well past nightfall when the last of the raiders were finally on Yankee soil. Sergeant Henry Stone later expressed the feelings prevalent among Morgan's men: "Wake up old Hoosier now. We intend to live off the Yanks hereafter and let the North feel like the South has felt of some of the horrors of war—horses we expect to take whenever needed, forage and provisions also. . . . This will be the first opportunity of the Northern people seeing Morgan and they'll see enough."

On the morning of July 9, Morgan's ten regiments descended on Corydon, in southern Indiana, which was defended by some 500 militia. After Morgan's men had dispersed these, the best account of the fight was related by Edwin Wolfe, editor of the local newspaper: "The shells made the ugliest kind of music over our heads. This shelling operation, together with the fact our line was about to be flanked on both wings at the same time, made it necessary for the safety of our men that they should fall back. This they did, not in the best order, it is true, but with excellent speed."

Morgan, however, was shaken at Corydon when he was informed by an innkeeper's daughter of news from the previous week that the entire North had already learned. Robert E. Lee had been defeated in Pennsylvania and had lost a third of his army. The remnants were being pursued back to Virginia. Furthermore, Grant had finally taken Vicksburg and over 30,000 prisoners. Morgan was shown a newspaper that verified the woman's report. The Rebel commander received the news like a blow to the stomach. However now, more than ever, it seemed, the spirit of the South would depend on the continued success of his raid.

With the Pennsylvania option gone, the Rebels would keep to a specific plan that had been outlined in advance: head west across Indiana to Ohio, describe an arc of destruction above Cincinnati, and then hit the Ohio River again at a crossing to Virginia. (On June 20, the western counties of Virginia had seceded from their state, forming West Virginia, which was loyal to the Union. This was another piece of bad news that Morgan learned at Corydon.)

The state of Indiana, meanwhile, had been aroused to the point of panic. Thanks to Ellsworth, Governor Oliver P. Morton was preparing for an assault by the raiders against Indianapolis, including the liberation of 6,000 Confederate prisoners there who would be enlisted to join Morgan's "ten, or twenty," thousand men. Nathan Bedford Forrest was reported to be following Morgan with his own division. A state of emergency was declared, and rail lines were commandeered to send militia south; Illinois was called upon to add volunteers.

Morgan's raiders, meanwhile, were heading east, to Ohio. On July 10, Morgan's men entered Salem, a particularly affluent and attractive small town, and proceeded to ransack it thoroughly, requisitioning horses and looting the stores. Raiders rode off with bolts of calico and other dry goods—one with a collection of ice skates, another with a canary cage containing three singing birds.

Arriving at the town of Vernon, scouts reported that 2,000 militia were barricaded in the city's streets. Morgan sent in a demand for their surrender. The militia, expecting 1,200 reinforcements to arrive shortly by train, carefully considered the offer but finally refused. Morgan's men were already moving on, but he repeated his demand that the militia surrender. Once again, the Indianans pondered and debated while the rest of Morgan's command pushed east. Finally, it was only the Confederate rearguard, consisting of one company, that proffered the final surrender demand. The militia, along with its reinforcements, bravely decided that their time to fight had come, before they realized that Morgan had already gone.

On the 12th, Morgan's men found a meat-packing plant that contained 2,000 smoked hams. If the uniforms and appearance of Morgan's men were not exactly identical—usually featuring broad sombreros, pants tucked into boots, long hair and gray or butternut clothing—the command became unified by the feature of a ham on each saddle, cooked gradually along the way.

The strength of the militias in Indiana and Ohio was probably a greater problem for Morgan than he had anticipated, since his previous experience in Kentucky and Tennessee had been with communities containing far fewer able-bodied men not already committed to the war. (Grierson, in Mississippi, had encountered hardly any militia at all.) Also, although Morgan's 2,400 men were capable of demolishing any one militia unit, the Confederates did not ride as a united command. Needing to live off the land, they generally took parallel roads, and entire companies were constantly dispatched to right and left as scouts, foragers and diversions. If 500 militia gathered to defend a town, they would, for example, be confronted by only 200 of Morgan's men, an advance guard of one prong of his advance.

Nevertheless, the Southern riders could converge to disperse any particular militia that threatened their progress. By July 12, however, a greater menace was already apparent in their rear: regular Union cavalry were behind them and closing fast.

Edward Hobson's division of 4,000 men, including the cavalry of Frank "Meat Axe" Wolford and James Shackleford, had followed Morgan across Kentucky, crossed the Ohio, and was now on his tail. While Morgan waded through the untrained formations in front of him, his nemesis was just 12 hours behind. The Union cavalry had one disadvantage of following in Morgan's wake: the Rebels had picked the countryside clean of most of the good horses. However, some Union citizens, having successfully hidden their horses from the invaders, brought them out for the benefit of the Federal pursuit. The Union cavalry also marched to a different tune: while Morgan's men rode to the cacophonous sound of church bells ringing the alarm, the Yankee horsemen were serenaded by rows of people singing "Rally 'round the flag boys" as they pursued the Confederate raiders. ("Rally Around the Flag" became such a theme song for the Union chase that a Federal officer, in later years, claimed he couldn't hear water dripping without the words of that song coming to mind.)

On July 13, Morgan entered Ohio, prompting that state to declare a state of emergency. General Ambrose Burnside, he of Antietam hesitation and Fredericksburg slaughter, was now commander of the Ohio Department, based at Cincinnati. He was supposed to be moving an army south to join with Rosecrans for an advance against Bragg, but John Hunt Morgan's invasion of his domain was a challenge that Burnside first had to meet.

Morgan had meanwhile dismissed the suggestion that he take Cincinnati. Sam Taylor had come back from that city to report martial law had been declared and business had been suspended. There was no point in a street battle against what would no doubt be a huge militia; also, any delay would result in Hobson's cavalry catching up. The Confederates rode through the suburbs on July 14, continuing through the night and into the 15th, covering 90 miles in 35 hours. Along the way, they destroyed bridges over the Whitewater and Miami rivers and also derailed a train. "The train shot past us like a blazing meteor," according to Confederate Lieutenant Peddicord, "and the next thing we saw was a dense cloud of steam above which flew large timbers. Our next sight startled our nerves, for there lay the

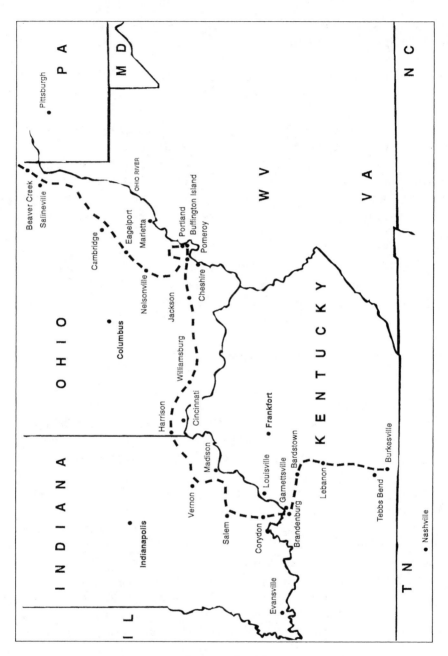

Morgan's raid across the Ohio River, 1863.

monster floundering in the field like a fish out of water, with nothing but the tender attached. Her coupling must have broken, for the passenger carriages and express were still on the track . . . Over three hundred raw recruits were on board . . . they came tumbling and rolling out in every way imaginable . . . all submitted without a single shot." Arriving 28 miles east of the city, the raiders found themselves unmolested outside Williamsburg, where they all went to sleep.

Morgan's command had by now suffered attrition, more through weariness than casualties, and was down to about 2,000 men. Pursuing Federal cavalry, coming upon Confederate raiders asleep in fields, would have to kick or bayonet them simply to wake them up. Many of the Rebels' horses had also given out, forcing their riders to drop out of the columns. Hobson, Shackleford and Wolford were now in pursuit directly behind the Confederates, while Henry Judah's cavalry had been ferried up the Ohio and disembarked at Cincinnati. Morgan thus had 10,000 Union regulars on his tail in addition to innumerable Ohio militia in front.

About the latter, Confederate James McCreary wrote, "The enemy are now pressing us from all sides and the woods swarm with militia. We capture hundreds of prisoners, but, a parole being null, we can only sweep them as chaff out of our way." The Ohio militiamen, when in small groups, were well advised to aim high, if they chose to fire at the Rebel raiders at all, so as not to invite a violent response. When gathered by the hundreds, however, they presented formidable obstacles to Morgan's men, who by now simply wanted to get out of the North.

It was on July 18, just as the raiders' hopes seemed to be at their lowest ebb, that the commander of the 2nd Brigade, Adam Johnson, encountered General John Morgan lounging on the porch of a store at an Ohio crossroads. Morgan was smiling and in good spirits. "All our troubles are now over," he said. The river is only twenty-five miles away, and tomorrow we will be on Southern soil."

The predesignated crossing point for the raiders was Buffington Island, where the Ohio River ran shallow, with sandbars in midstream. If Morgan's cavalry had, in fact, crossed over to Virginia at this spot, the raid itself would have been heralded as one of the outstanding Confederate exploits of the war. In response to the disasters of Gettysburg and Vicksburg, Southern cavalry would have run a circle around Cincinnati, even as New York City buried its dead from the anti-war draft riots that had commenced on July 13. But the Rebels in Ohio were not to be so fortunate.

On the afternoon of July 18, Morgan ordered his command to pause at the town of Chester, while the long columns regrouped, and to make sure everyone was accounted for. When the raiders got to Buffington Island, it was after dark and scouts reported the crossing defended by 300 Union troops, with two cannon. His command intact, and with Southern territory in sight, Morgan made a hard decision: he would not attack in order to force a nocturnal recrossing of the Ohio; he would wait for the next morning, the 19th of July.

At dawn on the 19th, two regiments of raiders advanced on the Union force defending the ford, only to find they had run away during the night, after tossing their cannon into the river. But shots were heard at the rear of the Rebel column—Judah had arrived. In a dense fog, some of Basil Duke's men fought back lead elements of the Union cavalry, killing or wounding twenty and capturing fifty more, almost including Judah himself. But the Union brigade was nearly double the size of Morgan's entire command; Federal regiments came pouring onto the field as Duke alternately counterattacked and fell back, his men firing from behind trees.

Then firing was heard on the Rebel right as pickets came racing in to report Edward Hobson's brigade had also arrived. Hobson's cavalry, under Shackleford and Wolford, had followed Morgan's men all the way across Indiana and Ohio and now had them pinned against both the river and Judah. The Ohio, meanwhile, had been swollen by recent rainfall; it was, in fact, accessible to gunboats and three of these now sat astride the ford. Perhaps 30 Rebels got across before the gunboats approached; another group of raiders was able to cross on an improvised flatboat. The main force, however, was trapped.

Federal artillery, from three sides, crashed into the raiders' position, which became a charnel house of overturned wagons, dead horses and fleeing men. Duke rallied his men against Judah while Johnson's brigade tried to hold off Hobson; the two timberclads and one ironclad gunboat, armed with 24-inch Dahlgrens, enfiladed the Confederate force from the rear. Assailed by a vicious crossfire and outnumbered four to one, John Morgan had only one remaining option: breakout. There was still a small gap between Judah's and Hobson's converging brigades. Morgan assembled the bulk of his command, and while Duke and Johnson held the flanks he raced out of the trap, between the converging Union cavalry and away from the gunboats. Adam Johnson was able to disengage and follow the retreat as a rearguard. The raiders suffered 120 dead or wounded at Buffington Island while Duke and 700 Rebels laid down their arms; Morgan and 1,100 others escaped.

Less than 20 miles upriver, near the town of Portland, Morgan again tried to transfer his command onto Southern soil. Over 300 had crossed (including Johnson and Ellsworth) when gunboats once again arrived on the scene. Morgan himself had been halfway across the river when they came up and could easily have escaped. However, he turned around and swam his horse back to the Ohio shore to rejoin the majority of his men.

The last week of Morgan's incursion into the North could not be described so much as a raid but as a determined band of fugitives endeavoring to escape capture. Morgan rode north, the Federal cavalry, as always, never able to get ahead of him, only to nip at his heels. On July 26, however, Morgan and his men, now reduced to about 300, paused to watch the dust clouds on either side of them advance to their front. The Confederates' horses were exhausted, as was their ammunition. They were surrounded, cut off, outnumbered and outgunned.

Morgan, then a day's ride from Lake Erie, due northwest of Pittsburgh, craftily surrendered to a militia captain he had just captured, William Burbick, on the condition that the Confederate and his officers all be paroled. Burbick agreed to the famous Rebel's request; however, the Union cavalry officers who soon came on the scene negated the militiaman's promise. Shackleford, in fact, became abusive toward Morgan until Kentuckian "Meat Axe" Wolford intervened. In any case, John Hunt Morgan was not paroled, or even placed in a prisoner camp, but was sent instead to the Ohio peniteniary in Columbus, where he was soon joined by Duke and most of his officers.

The "great raid" into Indiana and Ohio, which, excepting Lee's Gettysburg campaign, was the most dramatic incursion into the North during the Civil War, has become a subject of some debate. With barely 2,500 men, John Hunt Morgan caused the mobilization of 55,000 Indiana militia, 65,000 in Ohio, and tied down 10,000 Union cavalry that would otherwise have been occupied in Kentucky and Tennessee. Burnside was delayed in going south to join Rosecrans and, on September 20, Braxton Bragg decisively defeated Rosecrans' army for the South's greatest victory in the western theater: Chickamauga.

On the other hand, in the "great raid," John Morgan had taken one of the South's two or three best cavalry formations and, against orders, ridden it into oblivion. His idea of joining hands with Robert E. Lee's army in July 1863 turned out to be nothing but a dream.

In a dramatic episode that caused severe recriminations in the North and prompted celebration in the South, Morgan and six of his officers were able to tunnel out of the Ohio penitentiary and escape in November 1863 (after someone had given scout Tom Hines a copy of *Les Miserables*). Then, evading the threat of court-martial for his disobedience of orders, Morgan succeeded in gaining from Richmond a new command that included many of his veterans. Operating out of western Virginia, however, he was never able to duplicate his earlier success.

In November 1864, Morgan was surprised at Greeneville, Tennessee, and shot down by Union cavalry. His last words may have been "Don't shoot. I surrender." The last words he heard may have been "I know who you are, you damned horse thief." Like Hector of Troy, his body was paraded among his enemies and mistreated. Finally, Union General Alvin Gillem interceded and arranged for his body's proper treatment.

Although John Hunt Morgan died a less than glorious death, and his career as a "knight savior" of the Confederacy was extinguished midway through the war (somewhere in the north of Ohio), his legacy continued after he had long since been put in the ground. Several hundred of Morgan's command had gotten back from the raid into the North; a few hundred more had been left behind; still others returned in prisoner exchanges.

Anyone who reads extensively about the last year of the Civil War encounters, time and again, a term that describes a certain kind of Rebel soldier. These are grim, hard fighters—cavalry—who, in the accounts of Nathan Forrest, Joseph Wheeler, Jefferson Davis and others, are described to us as elite troops who were a source of pride for the Confederates wherever they fought in those last, desperate months of the war. The term that describes these soldiers is "Morgan's men."

7

Custer at the Washita

BY STEPHEN TANNER

During the late 1860s, the military situation on the Great Plains was in flux, a situation due not only to the strength of the Indian tribes who held that territory, but to the inhospitable nature of the Plains themselves. Essentially, the tidal wave of white immigration that had inexorably claimed the East, producing time and again entire armies of citizen militias to fight for New York, Kentucky, Illinois and other present-day states, simply stopped short once the frontier had reached the Plains. That vast stretch of the American continent, roughly from Texas to the Canadian border, and Kansas to the Rockies, was (and still is) hostile to large-scale settlement, and a fight for the territory was not a priority for anyone save those few intrepid individuals willing to live on it.

That whites were streaming through the Plains at all was primarily a result of the repeated discoveries of gold in the West—first, and most famously, in California in 1849. The tiny but persistent streams of immigrants in wagon trains, intently heading for the few passes in the Rockies then known to exist, succeeded in familiarizing the Western Indians with white America, and at the same time prompted the government to provide some degree of protection for the vulnerable immigrants. The "war" for the territory, in any case, was slow to get underway and the U.S. government had initial difficulties in projecting force there.

Gold was also discovered in Colorado in 1858, prompting another wave of invaders, some 100,000 to the area around Pike's Peak and that resulted in the establishment of Denver. In 1865, the metal was again found, this time in Montana, and another white avenue through the forbidding territory was established as the Bozeman Trail.

In addition to the gold-seekers, the population of the United

States—some 40 million by 1865—was also seeping westward as farmers and ranchers, although these did not comprise large numbers. Other individuals were attracted by the gigantic buffalo herds, a phenomenon found nowhere else in the world, that in their millions roamed back and forth across the vast, empty territory. Yet, despite an increasing number of white incursions into the endless Western spaces—in most cases simply "passing through"—the Indians still ruled the Plains.

The Plains tribes of the mid-1800s were most comparable in history to the steppe people of central Asia centuries prior to their unification into "empires," first by the Scythians and ultimately by the Mongols. Being nomads, they considered huge swathes of territory to be their "home," resulting in perpetual warfare with neighboring tribes that had, over the centuries, become ingrained in their culture. Given the Plains Indians' lack of literacy or an alphabet, fixed communities, or any industry beyond Stone Age in sophistication, excelling in battle was the primary means for a young man to achieve social status. Taking scalps or "counting coup," not to mention killing enemies, regardless of specific political or territorial imperatives, were sure-fire guarantees that a warrior could advance his station. The sudden presence of whites in Indian territory offered new opportunities to demonstrate prowess.

Although the first whites encountered by the Indians west of the Mississippi were no doubt formidable characters, after 1849 the average toughness of the invaders began to decline in direct proportion to their numbers. If approached with sufficient cunning, whites proved easier to kill than fellow Indians, and, although the evidence is scanty, many may have, reflexively, tried to surrender at the very climax of a brave's attack. The Plains warriors were fantastic horsemen, and rode animals born and bred in the territory. They were fierce hand-to-hand fighters and effective bowmen; moreover, after the government, in a controversial approach to "pacification," began providing firearms and steel knives and hatchets to the tribes (allowing traders to provide still more) the Indians' military capability moved up another notch.

In warfare, the Indians' only weakness was in their system of command, since their operations lacked its twofold raison d'être: coordinating the men, and compelling them to perform tasks they would not ordinarily undertake on their own (for example, attacking at a certain point and time). Indian battles were generally wild, swirling melees,

and, when taking on 20 or more whites, the Indians invariably needed a huge numerical superiority to succeed, to make up for the fact
that most warriors simply fought as they pleased.

Indian fighters hated to die or be wounded. This may sound odd,
but the truth is that when serving, presumably with obedience, under
a system of rigid command, a soldier's reluctance to be shot is less
important than the orders of his superior officer. The Indians, on the
other hand, employed individual common sense in their fighting in
order not to gain a ghastly wound. Their specialty was hit-and-run;
then hit again later—always when the enemy was vulnerable. The one
battlefield tactic they were able to execute with precision was the
ambush. As for the robotic, blue-clad formations that came prome-
nading onto the Plains after 1865, responding to bugle signals and firing in unison, these were simply to be avoided. After a hundred miles,
or two, the men who comprised these large units would soon be found
singly or in small groups, as couriers, stragglers or deserters, and then
they were not quite as fearsome. In one-on-one combat, according to
nearly all reports, the Indian warrior had no superior.

By the mid-1860s, those whites who traveled west were doing so
only by virtue of a large quotient of courage, in addition to their
apparent optimism and/or greed, and American citizens were being
killed randomly, apparently at will, by warriors who were still able to
make certain that the bulk of the Great Plains remained "Indian territory."

Earlier, the Plains Indians had suffered two travails that echoed
what their ethnic cousins had endured in the East: running afoul of
white centers of population. In 1862, Little Crow led his branch of the
Sioux in an offensive in Minnesota, succeeding in killing over 800
white civilians and soldiers. American troops and militia, under
General Alfred Sully, however, defeated his tribe and then hanged over
30 warriors, the remnants of the tribe traveling west to join their fel-
lows in the Dakotas. In 1864, Indian successes against individual
whites and stagecoaches in Colorado enraged the population around
the newly founded city of Denver, and a militia was formed to seek
retribution. At Sand Creek this militia, which closely resembled an
angry mob, destroyed Black Kettle's Cheyenne village, in the process
failing to distinguish between men, women and children. Although
celebrated in Denver itself, news of the mass killing, as well as the
mutilations of Indian corpses, sickened the East, and the U.S. govern-

ment, including the Commissioner of Indian Affairs, condemned the action. The government even offered compensation to the surviving Cheyenne who had lost relatives in the attack, although this must have seemed a grim curiosity to the recipients: the compensation was 160 acres of land.

The Sand Creek massacre, tragic as it was, provided a fresh impetus for Plains Indian warriors, and during the following summer the Oregon Trail was repeatedly plundered in revenge. In the summer of 1865, General Patrick Conner attempted a sweep of the territory with a large force of soldiers but was unable to force a confrontation with the elusive war parties.

Following the discovery of gold in Montana, whites began traipsing the Bozeman Trail in a veritable "invasion" of Indian-held land. The Sioux chief Red Cloud, however, rampaged throughout the summer of 1866 and effectively closed the trail. Whites had meanwhile set up forts deep inside Indian territory, although these became virtual prisons for the garrisons contained within. Not only was the surrounding country forbidding, but small groups of soldiers venturing out of the forts were often attacked by braves who struck without warning. In December 1866, an officer named Fetterman rode out of Fort Phil Kearney to rescue a wood train with 80 men, a force he considered sufficient to "ride through the whole Sioux nation." His entire command was wiped out, even as the wagon train he was rescuing took advantage of the commotion to get its wood back to the fort.

In the summer of 1867, the Hayfield fight, followed the next day by the Wagon Box fight, took place, in which soldiers were surrounded and attacked by superior numbers of Indians. The Wagon Box defenders were saved by cavalry charging to their rescue from Fort Phil Kearney. The Hayfield troopers (or at least half of them), however, were forced to survive the onslaught on their own, since the soldiers in nearby Fort C.F. Smith had been reluctant to come out of their stockade.

In the fall of 1867, Red Cloud agreed to a truce which stipulated that the whites would close down the Bozeman Trail and abandon its nearby forts. The government asked in return, by way of "fine print," only that Red Cloud compel his people to behave peaceably, learn to farm and send their children to school. The treaty, not clearly understood by most of the Indians, was signed with many "X's." But some Sioux, notably those led by Crazy Horse and Sitting Bull, would have nothing to do with it.

To the south, Cheyenne, Kiowa, Arapaho and Commanche braves continued to prey on whites along the Platte River route to Colorado and California. The great problem for U.S. punitive expeditions, however, was simply finding the Indians in order to give battle. Entire armies would trample the Plains without seeing an Indian, or a sign of one, even as reports of additional slayings and massacres from other locales continued to come in. Meanwhile, many Indians were readily accessible on designated "Reservations," particularly in winter, drawing from government supplies and professing friendship. It soon became clear that certain tribes had accommodated to the white demand—including its trade-off in sustenance and supplies—that the Indians give up their nomadic ways, but others had not. The army attempted to hunt down the "hostiles," even as the Interior Department fed and supplied the "friendlies." Inevitably, the two groups became mixed, the number of friendlies increasing dramatically during winter. In any event, although all the Plains Indians were feeling some pressure to come to terms with white demands, a majority of young Indian men in the recalcitrant tribes were still either avowedly hostile, or at least hostile in good weather. And no one could catch them.

While the Interior and War Departments pursued their contradictory policies in order to solve the problem on the Great Plains, American public opinion was in itself deeply divided on the Indian question. If admiration for the "noble redman" seems pronounced today, particularly for those Native Americans with feather bonnets on painted ponies roaming freely on the plains—as opposed to, perhaps, the Mohawk who suddenly appeared out of the forest at the cabin window in upstate New York in the previous century—in the 1860s this was no less the case. In the East, public sympathy was largely attached to the Indians as romantic figures, and those homesteaders attempting to encroach on their territory—by definition people who were not successful elsewhere—had no constituency beyond their fellow Westerners and the oft-humiliated military. The killing of Indians was not highly regarded in the East (as it nevertheless was in the West), so "Indian fighters" were subject to intense public scrutiny.

As for the U.S. armed forces, after April 8, 1865 the re-United States possessed the largest and most experienced pool of professional military men in the world, despite some 600,000 recent deaths in battle. Among the cavalry, the Southern arm, including men like

Nathan Bedford Forrest (fortunately for many) had retired from the profession of conducting mobile warfare. From the Northern cavalry, however, great talents had emerged, as well as a plethora of other officers who at least had been hardened in battle. Among the Union cavalrymen, no one cut as dashing and charismatic a figure as the "boy general," George Custer.

If Custer had never fought Indians, he would still be a major historical figure for his exploits during the Civil War, during which he proved to be a brave, skilled commander of cavalry—and also lucky. (Napoleon once commented that, given a choice, he would prefer a general who was lucky rather than good. During the Civil War, in fact, Custer was both.) Rushing to Bull Run, straight from his graduation from West Point, Custer immediately distinguished himself under fire. This earned him a promotion to McClellan's staff, roughly from the Peninsula campaign through Antietam.

In the first two years of the war the cavalry of the Army of Northern Virginia, under J.E.B. Stuart, had (literally) run rings around the Union horsemen and, as a result, just prior to Gettysburg, it was decided to energize the Union cavalry by elevating three youngbloods to brigade command. Eton Farnsworth would be killed within days, but 27-year-old Wesley Merritt went on to become a highly-respected commander. The third new brigadier general, Custer, who leaped from brevet captain to general at the age of 23, would become the soundest choice of all. It appears, in fact, that at the time of his promotion he had already prepared a custom-made general's uniform: black velvet from head to toe and decorated with stars.

On July 3, 1863, "Jeb" Stuart had finally joined the huge battle in Pennsylvania and was aiming his unsurpassed horse army toward the rear of Cemetery Ridge, even as 12,000 Confederate infantry attacked from the front. As part of Judson Kilpatrick's division, Custer led his Michigan Brigade in a fierce charge against Stuart's men, only to encounter a stone wall that disrupted his attack. His men were badly cut up, but Custer reformed the brigade. Again and again they charged. Stuart launched six regiments against Custer's left, but the new general, everywhere at the head of his troops, retreated, reformed, and charged again. Custer had two horses shot from under him during the day's fighting, and his brigade lost 219 men, but Stuart never got through to the Union rear. The "boy general" had validated his promotion.

The following October the largest cavalry battle of the war was fought at Brandy Station in Virginia, and, although Stuart was not defeated, the quality gap between Union and Confederate cavalry was seen to have closed. Custer enhanced his reputation for decisiveness and gallantry, again losing two of his horses in the fighting.

At Buckland, Custer's Michigan Brigade lost 214 men, including an entire battalion captured, in what became a dismal defeat for Kilpatrick's division. Privately, at least, Custer turned on his commander, who was commonly nicknamed "Kill Cavalry," and blamed Kilpatrick's staff for erroneously placing his lost battalion. Nevertheless, the high regard for Custer increased, especially among his own command. A Michigan officer, James Kidd, wrote home that "Custer was always on horseback. He never was seen on foot in battle, even when every other officer and man in his command was dismounted. And he rode close to the very front line, fearless and resolute." Between his flamboyant uniforms, his flowing blond hair, and his refusal to make himself inconspicuous in battle, even when the rest of his men were on foot or behind cover, it's amazing that Custer survived the war at all.

In the spring of 1864, U.S. Grant came east to take command of all the Union armies, and he, in turn, appointed Phil Sheridan to take charge of the cavalry corps of the Army of the Potomac. The dynamic Sheridan quickly appreciated the resource he had in Custer, and the two men began a friendship that would last beyond the war. In May, the cavalry operated south of Grant's offensive that set down at the Wilderness, Spotsylvania Courthouse and Cold Harbor, and Custer found himself attacked by his friend and ex-West Point classmate, Rebel general Tom Rosser. On May 11, Sheridan's corps found Stuart at Yellow Tavern with only part of his force assembled. In the ensuing battle, one of Custer's men put a fatal pistol shot into Stuart, ending the Confederate cavalryman's legendary career, although he was capably succeeded by the talented Wade Hampton.

In the summer of 1864, Sheridan, largely in response to Jubal Early's surprise visit to Washington, D.C. with Stonewall Jackson's old corps, was ordered to clear out the Shenandoah Valley. Custer played a leading role in a series of victories, including Winchester and Fisher's Hill. At Toms Brook, Custer broke the Rebel line held by his old friend Rosser, thus beginning the "Woodstock Races."

On October 25, Custer was promoted to major general and given command of the Third Division, while his fellow prodigy, Merritt, was promoted to command of the First. At Cedar Creek, Rosser occupied Custer while Jubal Early launched a surprise dawn attack that soon had Sheridan's entire army fleeing from the field. Sheridan arrived at the battle in the afternoon, however, in an inspirational manner, and stopped the route. During the ensuing counterattack, Custer's right hook got through the Rebel army in a classic cavalry maneuver, and reduced it to a shambles. Custer thought he had taken thousands of prisoners, but it turned out that many Confederates surrendered more than once—to each of Custer's charging men they encountered while prudently vacating the field.

The Appomattox campaign became one headlong rush for Custer, including Saylor's Creek, where a third of the Army of Northern Virginia surrendered to his men, although it was muttered among other Union officers that Custer's division had not so much caused the surrender as been quicker to accept it. After Robert E. Lee's capitulation, nevertheless, Sheridan bought the table that Grant had used to write out his terms (for $20) and sent it to Custer's wife, Libbie, as a present. He included a note that said, in reference to the North's ultimate victory, ". . . permit me to say, Madam, that there is scarcely an individual in our service who has contributed more to bring this about than your very gallant husband."

That Custer was by nature an adventurer, and a hunter, and a man whose greatest moments in life, aside, perhaps, from some of those spent with Libbie, came through leading men into battle resulted in his continuing in the "peacetime" army, which was then struggling for control of the Great Plains.

One can only imagine the "culture shock" of Union officers, fresh from direct, confrontational-style warfare involving hundreds or thousands of men on a side, buttressed by ideology and patriotism, who subsequently found themselves Indian fighters. In Custer's case, as a man with great stamina who loved the outdoors, as well as any kind of challenge, the shift to the West was apparently invigorating, at least as opposed to occupation duty in the South. But the frontier army was comprised of a different breed of men from those with whom the young general had made his reputation. As in other cases when an advanced army has attempted to cope with guerrilla warfare—which was what the Indians practiced, albeit a picturesque version—the morale of the army was dismal, desertion rife, and patriotism a nonfactor. Custer's first attempt at a "search and destroy" mission with his new troops was disastrous.

In June 1867, Custer led the newly formed 7th Cavalry to seek out Indian war parties that had been terrorizing white settlers, wagon trains and stagecoaches along the Kansas–Nebraska border. He wasn't able to find the Indians, but immediately established a reputation among his men as a martinet: at worst, a pompous and vicious disciplinarian; at best a self-indulgent commander who brought along his own (female) cook and placed a priority on private pleasures, including frequent game hunts.

Then 27 years old, Custer may have been conflicted during this period between how he thought an elite unit should perform, including the enthusiasm they should bring to an operation, and the truth about his desultory, rebellious troops and their fruitless mission. In any case, his first venture onto the Plains resulted in his court martial conviction.

In 1867, Custer longed for his wife, and suggested in a letter that she head for Fort Wallace, which was near where his regiment was hunting hostiles. A wagon train of supplies coming out from Wallace, that he thought might also contain Libbie, was attacked by some 300 Sioux and Cheyenne, the 48 cavalrymen managing to hold them off. Meanwhile, Lt. Lyman Kidder, with ten men, had tried to reach Custer from Fort Sedgwick, but they were found by Indians and, after a running fight, massacred near a creek. Custer discovered their bodies, and later recorded after a minute inspection how horribly mutilated they were—a first-person description that rings hauntingly ironic in the light of events yet to come.

Desertions from Custer's command were frequent, one night over 30 men riding off. The next day, a dozen men simply deserted in daylight, and in full view of the rest of the regiment. Custer ordered these men to be overtaken and shot. Major George Elliott led the pursuit and his men shot three of the deserters, bringing them back to camp, wounded. Custer announced that they should not receive medical attention (though privately, to the doctor, he asked that they be seen to). One man subsequently died.

What really got Custer in trouble was when he returned to Fort Wallace, which had been attacked twice in his absence, and then decided that, since Libbie wasn't there, he would personally proceed to Fort Riley, where he could catch a train to achieve a marital reunion. In effect, Custer himself deserted his own command. In the march to Riley, with an escort of 70 men, Custer drove the troops so hard that some of them dropped back—or, "without authority halted," as Custer put it in his own account. Six men went back to find the stragglers and two of them were shot by Indians. Instead of dealing with the Indians, Custer kept on going, although one of the men was found to be only wounded. Custer had abandoned a trooper in his single-minded rush to reunite with Libbie.

The subsequent court-martial resulted in Custer's suspension from the army for a period of one year.

Throughout the spring and summer of 1868, no satisfactory battles were brought to bear against the proficient Indian warriors, many of them now armed with late-model guns, and American policy seemed increasingly impotent. In July, Cheyennes cut a bloody swathe through homesteads along the Saline and Solomon rivers in southern Kansas; in September a select group of elite scouts lost half its number after being trapped by Sioux and Arapahoes at Beecher's Island in eastern Colorado. The tally for the first half of the year was: 157 dead whites, 57 wounded, 14 women "outraged," and 1 man, 24 children and 4 women captured. The confirmed figure for Indian dead was eleven (though some say only two), with one wounded. Still, the majority of public opinion in the East, as well as the (corruption-wracked) Interior Department, came down basically in support of the tribes.

The War Department was simply frustrated. The key to the final government action, that resulted in the raid on the Washita, was the fact that the man who became president in November 1868 was named Ulysses S. Grant; the commander of the army was William Tecumseh Sherman; and the commander of the military Department of the Missouri was Philip H. Sheridan. None of these men had ever acquired a taste for defeat, and all three of them knew that a good army could operate in all-weather campaigns. If the Indians couldn't be tracked down in summer, they could surely be sought out in the winter. The hostiles would be found and fought, and they all knew the man who could do the job.

Custer, in his interesting memoir, *My Life on the Plains*, passes over his court-martial exile in euphemistic fashion: "While [the 7th Cavalry] were attempting to kill Indians, I was studying the problem of how to kill time in the most agreeable manner. My campaign was a decided success. I established my base of operations in a most beauti- ful little town on the western shores of Lake Erie . . ." However, when the telegram came from Sheridan: "Generals Sully and myself, and nearly all the officers of your regiment have asked for you, and I hope the application will be successful. Can you come at once?" Custer rejoined the army, two months ahead of schedule, eager to receive his orders. Sheridan, after consulting with Sherman, had determined on a winter campaign to devastate the hostiles. In a subsequent letter to Custer, Sheridan wrote: "Custer, I rely everything upon you and shall send you on this expedition without giving you any orders, leaving you to act entirely upon your judgment." Sheridan knew that Custer would perform as he always had, bringing optimum results from his command.

On the morning of November 23, 1868, the men of Custer's 7th Cavalry awoke near their base at Camp Supply, in Oklahoma Ter- ritory, to a raging blizzard. There was no specific time urgency to their mission; nevertheless, as if to emphasize his determination to over- come all obstacles, Custer ordered the men to move out through the storm. Since being recalled from civilian life on September 24, Custer had reimpressed his stamp on the regiment, color coordinating the horses within each company, forming an elite company of sharp- shooters, hiring Indian scouts (from the Osage tribe) and putting together a musical band. One of Custer's characteristics as a military leader was to ensure that his commands not only fought, but also looked and sounded, like elite troops. (During the Civil War his men had charged to the sound of the band striking up "Yankee Doodle.")

The blizzard abated after the first day and the 7th was left to deal only with the excruciating cold, the tramping through snow, and the difficulty of finding any sign of the enemy on the huge, trackless plains. They were heading south, from where reports had arrived of a Cheyenne encampment on the Washita River. On the fourth day, Custer dispatched Major George Elliott with three companies to ride ahead of the regiment to look for signs of Indians. At 1:00 on the 27th, a messenger from the advance party came riding into camp with the news that Elliott had come across the trail of a war party of about 100 men, heading south. It was later learned that this war party had been raiding into Kansas and was Kiowa—at the time, the 7th Cavalry had no idea of exactly who they were tracking.

Custer immediately issued orders that Elliott was to pursue the trail and meantime the rest of the 7th, save the wagon train, would rush to join him. Alexander Hamilton's grandson, Lt. Louis Hamilton, was officer of the day and so was required, to his chagrin, to stay behind with eighty men to protect the wagons. He was able to find another officer, nearly incapacitated by snow blindness, however, to take his place and Custer then allowed young Hamilton to join the advancing forces.

After a few tense hours, detached from the wagons, and unable to locate Elliott, the main body of the 7th finally came upon the Indians' tracks, which then combined with Elliott's own, and soon the regiment was reunited. The combined cavalrymen, some 600 strong, continued their advance in the dark.

At 2:00 A.M., one of the Osage scouts reported that he smelled smoke and so he, with Custer, crept up on a ridge that overlooked the Washita River. They saw a large herd of animals, which Custer thought must be buffalo. The sound of a dog barking, however, and then a bell (indicating that the herd was ponies), and finally the clincher—a baby crying—confirmed that they had indeed found an Indian encampment.

The Indians found by the 7th Cavalry were the Cheyenne led by Black Kettle, and their village was the first in a series of winter camps along a twelve-mile stretch of the Washita that included Arapaho, Kiowa, Commanche, another branch of the Cheyenne, and even a roving band of Apaches. Earlier that evening Kiowas, from the war party that the cavalry had tracked, had ridden into Black Kettle's camp and warned of a large trail made by shod horses they had found out on the snowy plains. This report was scarcely believed, but Black Kettle resolved to move his people closer to the other camps at first opportunity. The previous week, on November 20, Black Kettle had visited Fort Cobb and beseeched its commander, Colonel William M. Hazen, for supplies and protection against army aggression. Hazen, considering the Cheyenne's summer attacks on settlers in Kansas, replied that Black Kettle would instead have to make his peace with Sheridan. The resources of Fort Cobb would be made available to the Kiowa and Commanche, but not the Cheyenne and Arapaho.

On the other side of a ridge which separated him from the village, Custer had no idea of the size of the Indian encampment, but trusted in the ability of the 7th to crack any opposition. His entire career had been built on the principle of aggressive tactics, and was the very rea-

1. Sir Francis Drake (1540–1596). After years of ravaging
Spanish possessions in the New World, he led a fleet from
England on a pre-emptive strike on the Spanish coast that
delayed the Armada for a year.

2. Francisco Zurbarán's painting *The Defence of Cádiz Against English Attack*. After Drake had destroyed most of the shipping in the port and gone away, the Spanish attempted to evaluate the raid as a defensive victory.

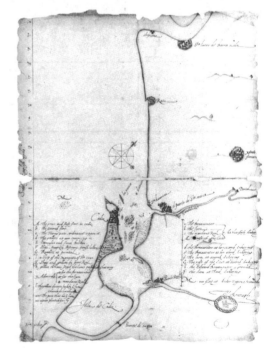

3. A map of the port of Cádiz drawn by William Borough, Drake's co-admiral during the raid.

4. French and Indian raiders on the New England frontier generally struck in good weather; however, nocturnal winter attacks as at Deerfield and Schenectady (depicted above) could catch settlements by surprise with devastating results.

5. "I have not yet begun to fight." The *Bon Homme Richard* grapples with the *Serapis* off Britain as the *Alliance* opens fire.

6. An idealized French engraving of John Paul Jones (1747–1792)
as a naval hero, replete with cherubim, "sea horses" (note the
webbed hooves), tridents, sea battles, cannon, flags and a globe.
Successful raiders could become folk heroes.

7. The Cossacks' triumphant entrance into Hamburg in 1813.
Many citizens turned out to welcome their liberators from the
French yoke.

8. Survivors of Napoleon's invasion of Russia had learned to fear Cossack attacks. Above, Prince Eugène's men fend off a raiding party outside Magdeburg, Germany.

9. Right, Czernichov was one of the most far-ranging Cossack commanders.

10. John Singleton Mosby. Though he disdained the use of
sabers in the field, the weapon was still standard issue for
encounters with photographers.

11. Mosby's raiders, having captured a wagon train, attempt to defend their capture against rescuing Union cavalry.

12. George "Lightning" Ellsworth, who, as John Morgan's telegrapher, pioneered the field of electronic warfare. According to which source is believed, his nickname derived from the speed of his fingers on the telegraph, his shiftlessness on all other occasions, or possibly to a fondness for a certain type of moonshine.

13. John Hunt Morgan (1825–1864), who dreamed of joining hands with the Army of Northern Virginia on Union soil. 14. Below, some of Morgan's men display the latest fashions in Confederate cavalry footwear.

15. George Armstrong Custer (1839–1876) made a name for himself as a dashing leader of Union cavalry in the Civil War, but he became even more famous afterward. He is shown here in the buckskin outfit he wore during the raid on the Washita.

16. Custer and the 7th Cavalry storm Black Kettle's village in a sudden dawn attack.

17. Koos de la Rey (1847–1914) was determined to resist the British Empire to the bitter end, winning the first and last victories of the Boer War.

18. General Paul Sanford Methuen (1845–1932) refused to quit the war until he atoned for his humiliation at Magersfontein. Unfortunately, he met de la Rey's commando again, in 1902.

19. Citizen-soldiers of the Boer republics, under the command of Koos de la Rey, celebrate their first victory of the war: the capture of the armored train *Mosquito*.

20. Lawrence of Arabia, standing center, with his bodyguard after the fall of Aqaba. Lawrence's understanding of the Arabs helped turn them into a cohesive force, and his raids tied up thousands of Turkish soldiers.

21. Thomas Edward Lawrence (1888–1935): scholar, writer, eccentric and raider.

son the army high command had sought him out for the task at hand. The decision was made to attack at dawn.

In keeping with cavalry tactics that had been validated time and again during the Civil War, Custer split his command, this time not just throwing out left and right wings to flank the enemy, but into four components, in order to completely surround the Indian village. Major Elliott, with three companies, was instructed to make the opposite side of the village to the left; Captain William Thompson with three companies was dispatched to get behind the village to the right, followed by Captain Edward Myers, who would take up an intermediate position between Thompson and Custer. Before setting off, Thompson nervously asked, "General, suppose we find more Indians there than we can handle?" Custer scoffed and replied, "All I am afraid of is we won't find half enough. There are not Indians enough in the country to whip the 7th Cavalry."

Not surprisingly, Thompson and Myers had difficulty finding their assigned positions in the dark, but Elliott was in place at daybreak. Custer had remained at the first point of contact with four companies, including the sharpshooters, plus the Osage scouts and the band. During the restless remainder of the night prior to the attack, Custer wandered among his command, at one point finding his Indian scouts speculating on the imminent battle. The Osage later told Custer they had assumed it would be a disaster for the cavalry. They thought that the flag bearer, being an important person, would be the man least likely to be exposed to danger, so resolved to take up position just behind him. If the battle went well they would join in, but if not, the Osage would hightail it, before the cavalry could trade them to the hostiles as a ransom for their extrication.

Just before dawn, as the troops, most of whom had snatched an hour or two's sleep, were assembling in formation, a strange phenomenon occurred. In Custer's words, "Directly beyond the crest of the hill which separated us from the village and in a line with the supposed location of the latter we saw rising slowly but perceptibly . . . and appearing in bold relief against the dark sky as a background something which we could only compare to a signal rocket, except that its motion was slow and regular. . . . The strange apparition in the heavens maintained its steady course upward. One anxious spectator, observing it apparently at a standstill, exclaimed: 'How long it hangs fire! why don't it explode?' . . . It had risen perhaps to the height of half a degree above the horizon . . . when, lo! the mystery was dispelled. Rising above the mystifying influences of the atmosphere, that which had appeared so suddenly before us and exciting our greatest apprehensions developed into the brightest and most beautiful of morning stars. Often since that memorable morning have I heard officers remind each other of the strange appearance which had so excited our anxiety and alarm."

Custer ordered the regiment's dogs to be killed once the village had been sighted, sparing only one of his own: a staghound named Blucher. Just prior to daylight he ordered his men to divest themselves of their overcoats and haversacks so as to be unencumbered in the fight to come. Six hundred men on horseback have difficulty sneaking up on anything during the still morning hours, however the sound of the cavalry steadily approaching was probably mixed with the ongoing rustling of the village's pony herd.

Custer had planned the attack to open with the band striking up the gay Irish tune "Garry Owen," but a hitch occurred when a shot suddenly rang out from the other side of the village, meaning the battle had already begun. Indian accounts claimed this shot was fired by Black Kettle himself, who had come out of his lodge and sensed the advance of troops, probably Elliott's, across the snow. On hearing the shot, Custer immediately whirled and ordered, "Give us Garry Owen." The band quickly got off a few bars, before the cold became a problem for their lips, their refrain in any event getting mixed with the bugles blowing "charge." Elliott's and Custer's shouting troops roared down into the encampment from opposite sides.

Indian warriors in the village made an amazingly quick recovery from their initial surprise, and the troopers met return fire immediately. Lt. Hamilton, who had been so determined to join the attacking forces, was shot from his horse, dead, in the first rush. Still, the cavalrymen were in and among the 50 Indian lodges within minutes, firing at everything that moved in the semi-darkness. Custer himself shot an Indian on his initial pass through the lodges, knocked over another with his horse and then took up position on a knoll from which he could better direct the battle. Scout Ben Clark later remembered, "I rode right beside Custer just ahead of the command. He would allow no one to get ahead of him."

While some warriors defended their families to the death in front of their lodges, others raced to the river to shoot back from behind its bank or took up firing positions in nearby ravines. Many women and children stayed inside their shelters but others attempted to flee, at great risk of being shot.

An Indian rushed up to Captain Albert Barnitz and both the officer and warrior dodged left and right, then fired simultaneously. Barnitz thought he got the Indian, but the warrior's point-blank shot from a Springfield rifle tore through Barnitz's midsection and knocked him from from his horse. The captain, to everyone's surprise, survived the wound. Custer's younger brother, Tom, a reckless fighter who often seemed anxious to "prove" himself to his older sibling, was wounded in the hand during the assault.

Captain Frederick Benteen, leading his squad through a wooded area next to the village, was confronted by an Indian he thought was only a boy, so he tried to make peace signs. The Indian fired at him with a pistol but missed. Benteen made more peace signs and the Indian fired again. The warrior's third shot hit Benteen's horse and then the captain finally fired back and hit the Cheyenne. The brave was a young man named Blue Horse, aged twenty-one.

Black Kettle, then 67 years old, was killed while trying to get away. He had jumped on a horse and pulled his wife up in front of him, but a trooper shot him as he attempted to cross the river; his wife was then killed beside him.

Custer himself admitted that civilians were bound to be killed in an attack such as this: ". . . orders had been given to prevent the killing of any but the fighting strength of the village; but in a struggle of this character it is impossible at all times to discriminate . . ." Private Delos G. Sandbertson was less delicate in recounting, "We fired whenever we could see a top-knot, and shot squaws—there were lots of them—just as quick as Indians. We just went in for wiping out the whole gang."

Custer's Osage scouts, having gained enthusiasm for the battle, seemed particularly indiscriminate in their attacks against women and children until Custer put a stop to their efforts. When the contingents of Thompson and Myers charged into the village, just after Custer and Elliott, and helpfully guided by the sounds of gunfire, the escape routes for the Indians were shut off. When the battle began, many Indians had gotten out through the gap to Elliott's left, in the direction of the other villages. Custer reprimanded Myers's men, who were chasing and firing at Indian women and children trying to get away.

Custer told Ben Clark, "Ride out there and give the officer command-ing my compliments and ask him to stop it."

Trooper Sandbertson left behind a fascinating account of his expe-rience in the battle that not only reveals something of the participation of the Indian women, but also what it feels like to be scalped. The trooper told a journalist:

> When it was full daylight, we all gave a big yell and charged right down into camp. The lodges were all standing yet, and lots of Indians in them. As we run through the alleys, a big red jumped out at me from behind a tent, and before I could short-en up enough to run him through with my bayonet, a squaw grabbed me around the legs and twisted me down. The camp was then full of men fighting, and everybody seemed yelling as loud as he could. When I fell, I went over backward, dropping my gun, and I had just got part way up again, the squaw yanked me by the hair, when the Indian clubbed my gun and struck me across the neck. He might just as well have run me through, but he wasn't used to the bayonet, or didn't think. The blow stunned me; it didn't hurt me the least, but gave me a numb feeling all over. I couldn't have got to my feet then if all alone, while the squaw kept screeching and pulling my hair out by the handful.
>
> I heard some of our boys shouting close by, and the squaw started and ran—one of the boys killing her not three rods off. The Indian stepped one foot on my chest, and with his hand gathered up the hair near the crown of my head. He wasn't very tender about it, but jerked my head this way and that, and pinched like Satan. My eyes were partially open, and I could see the beadwork and trimming on his leggings. Suddenly I felt the awfulest biting, cutting flesh go on round my head, and then it seemed to me just as if my whole head had been jerked clean off. I never felt such pain in all my life; why, it was like pulling your brains right out. I didn't know any more for two or three days, and then I came to find that I had the sorest head of any human being that ever lived.

The battle for the lodges lasted no more than fifteen minutes, although Indian warriors still held out along the river and in the trees.

Louis Hamilton had been killed, and fourteen other men wounded, but the Indians had gotten by far the worst of it: over a hundred slain.

Even as the 7th Cavalry occupied the village and the battle appeared won, the troopers were curious to notice that, instead of dwindling pockets of resistance, the number of Indian warriors in sight, just outside of rifle range, seemed to be increasing on all sides. From where Custer had attacked, Indians had come up and taken all the overcoats and packs that the men had been ordered to leave behind. The dog Blucher's body was later found in this area with an arrow in it. On the ridge lines surrounding the encampment, the troopers could look up to see evidence of growing Indian strength.

Custer had assembled his prisoners, 53 women and children, most of whom had stayed inside during the fighting, in lodges in the center of the village. When the shooting began to die down, he took his interpreter into one of the lodges to speak with Indian women in order to find exactly whom he had just attacked, and to ascertain what other strength was in the area. Black Kettle's sister, after bemoaning the fact that her brother's inability to control his young warriors had brought such a disaster to her people, informed Custer for the first time that the village was part of an entire series of Indian encampments along the Washita. The cavalry had attacked the easternmost, but there were others nearby. The 7th Cavalry, having overwhelmed the village, might itself now be overwhelmed.

After the village had been taken, and the greatest challenge had momentarily become those Indians getting away, Major Elliott had decided to pursue a fleeing group to the east. Calling for volunteers, he was joined by the regiment's sergeant major and 18 men. Elliott's career in the army had stalled and in this, the biggest confrontational battle yet forced against Indians since the Civil War (or at least since Fetterman), one can speculate that he saw an opportunity to achieve an exploit that would finally earn him a further promotion. Before riding off, he shouted to Lt. Owen Hale, "Here goes, for either a brevet or a coffin!"

Elliott and his men chased the fleeing Indians some two miles east before they were met by large numbers of warriors coming from the other camps. Cut off from the regiment, Elliott ordered his men to dismount in tall grass and form a circle for all-around defense. His position was unfortunately overlooked by high ground, however, and Indians simply fired into his command from behind cover. At one point in the fight, an Arapaho named Tobacco, armed only with a club, rode into the circle and struck three troopers before a cavalryman shot him down. The troopers were probably already decimated through fatalities and wounds before the Indians mounted their final rush that overran the position. Everyone was killed, and then Elliott's men continued to hold down warriors even after their death, as the victors of the engagement worked on the bodies.

An Indian later said that Sergeant Major Kennedy, one of the most respected soldiers in the regiment, was the last man standing, and the warriors had the notion of taking him prisoner. Kennedy, apparently under no such illusion that this would happen, or even be desirable, approached an Indian empty-handed, and then suddenly drew his sabre and thrust it through the brave. Kennedy was shot repeatedly, over 20 times, and his brains were, literally, then bashed out with tomahawks or clubs. His corpse was the only one of Elliott's command found outside the defensive circle.

Elliott's dilemma was unknown to the other men of the 7th, although Lt. Godfrey, who had also ridden out in that direction, came tearing back to report that large numbers of Indians were advancing from the east, and also that he had heard firing from where Elliott had gone. Around the village, the Indians on the surrounding heights were not attacking but their strength appeared to be growing. Captain Benteen, taking three troops, charged one group of close-in braves and drove them off. At one point, the regiment's quartermaster came barreling into the village with a small escort and an ammunition wagon, having run a gauntlet of warriors. Just after his arrival the Indians launched an attack from the direction he had come.

Meanwhile, in the lodge where Custer was interrogating Black Kettle's sister, she had joined his hand with that of a comely young Cheyenne woman and made a speech, which he could not understand but which in any case seemed to amuse Custer's interpreter. Custer finally asked what was going on, only to be informed he had just become "married" to the young woman. One of Custer's scouts volunteered the information that, upon his own marriage to an Indian, he had immediately been besieged for handouts from what turned out to be an enormous network of in-laws. In any case, Custer already had Libbie and firmly disavowed the ceremony, although in the future he would continue his acquaintance with the women present in the lodge, since they were now his prisoners.

With the village secured, the cavalrymen made a detailed inventory of its contents, including 573 buffalo robes, 210 axes, 140 hatchets, 35 revolvers, 47 rifles, 535 pounds of powder, 1,050 pounds of lead, 4,000 arrows and arrowheads, etc. The number of Indian fatalities was put at 103, Custer receiving the reports of his company commanders. He later claimed 103 "warriors" killed, although it is beyond a doubt that the figure included noncombatants. (The Cheyenne later claimed a dozen warriors lost, with twice that many women and children.) While searching the lodges, the soldiers found abundant evidence of prior Indian attacks against white settlers, including letters, pieces of clothing, daguerreotypes and other homesteader possessions.

As the afternoon progressed, the 7th's task became to destroy what Indian property they had seized, even while nervously assessing the growing number of warriors surrounding them. Foremost among the Indian property was the pony herd of over 800 animals. After instructing his prisoners to choose mounts for their ride back with the cavalry, Custer then ordered the destruction of the remaining ponies. No more ghastly example of army ruthlessness could have been imagined by observing warriors than the killing of some 775 Indian ponies in a field outside the village. At first the men tried to cut their throats, but the ponies had an aversion to the smell of whites, or at least to their perceived intention, and were difficult to approach. The troopers then resorted to shooting them. After choosing a well-made lodge to take back with him, Custer ordered the village itself burned to the ground, some of the teepees exploding, because of the stores of gunpowder contained within.

While Custer's attack on the Indian camp was no doubt the worst nightmare of most of its residents, one person whose heart may have soared, if only momentarily, at the brief sounds of "Garry Owen," the bugles blowing "charge" and the shouts and gunfire of the 7th Cavalry was Mrs. Clara Blinn, who, along with her infant son, was being held captive by the Indians at the time. The 23-year-old Mrs. Blinn, who had been taken from a wagon train in early October, was executed at point-blank range during the attack, however, then scalped, and her baby's skull was crushed. It would not be difficult to suppose that Indians, seeing their own women and children being shot during the battle, killed Mrs. Blinn and her child, William, in retaliation; however there may also have been a motive of "hiding the evidence." Clara, if interviewed by Custer, might have had derogatory reports to make about individual Indians.

It may have been the combination of motives that resulted in the immediate deaths of Mrs. Blinn and her son, and, at any rate, Custer and his men didn't even know they were there. Their bodies were discovered two weeks later when the 7th Cavalry, with Sheridan, revisited the battlefield. Controversy exists over exactly which village on the Washita Mrs. Blinn's and Willie's remains were found in. Some claim she was in Black Kettle's village, while others (including Custer) claim she was in Satanta's Kiowa encampment, five miles away, or even farther. After the battle, the Indians understandably had a motive to disavow her presence to inquiring whites, while Custer, likewise, would have had no reason to aver that a white woman and her baby were killed as a result of his attack. In any case, Clara Blinn was poignantly survived by a letter she had gotten out just prior to her death, which pleaded for someone to ransom or rescue her.

Another report from the battle, sometimes confused with the killing of William, is that a small white captive, about eight years old, was suddenly knifed to death by an Indian woman during the fight for the village. That a boy was killed in such a manner is indisputable, but it seems the boy was an Indian and the perpetrator may have been his mother. Hiding behind a knoll, the two were under fire from troopers when the woman suddenly grabbed the boy and thrust her knife through his chest in full view of their attackers. The woman was then shot in the forehead by a cavalryman. Since Custer was taking prisoners in this battle, and certainly had never demonstrated a wish to kill children, the squaw's action might seem rash; it's possible, however, that the woman had been exposed to white atrocities at Sand Creek, four years previous, and made an instinctive judgment on the child's best future.

After the pony herd had been annihilated, and the village itself razed, Custer was faced with a hard decision that called for the utmost prudence. Now aware that he faced a large concentration of Indians— some 6,000 in the string of villages along the Washita as it later turned out—his first priority was to reunite his command with its supplies, and before the warriors surrounding him discovered the location of his lightly defended wagon train, still ten miles back, sitting in the middle of a vast snow field. Without the train, and having just stirred up a veritable hornets' nest of Indians, the prospects for the 7th on the Washita would drastically dim.

Although he never lost his aggressive attitude, Custer had already experienced an encirclement battle during which doom stared him in the face: at Trevilian Station during the Civil War. On September 10, 1864, Phil Sheridan took on Wade Hampton's Confederate cavalry corps, and as part of his plan ordered Custer's Michigan Brigade to flank the Rebel right. Custer charged headlong through the underbrush and got into Hampton's rear, taking his train, along with his rear horses and horse-holders. Hampton threw in his reserve, however—Tom Rosser and the Laurel Brigade—and Custer was hard-pressed at his front and to his right. The real disaster occurred when Fitzhugh Lee's division, as Hampton had planned, came up from the left and severed Custer from the rest of Sheridan's command.

Surrounded, taking fire from all directions, and with Wolverines falling on every side, at one point Custer's flag bearer fell, mortally wounded. Custer tried to grab the staff but the dying man's grip failed to relinqish it. Custer tore the flag from its staff and draped it over his shoulder. Wesley Merritt's brigade had meanwhile been driving in Hampton's center, however, and Sheridan, seeing Custer's plight, soon organized an attack that got through to the Michiganers. Nevertheless, Custer had lost 416 men, along with half his officers, and had teetered on the brink of catastrophe. To add insult to injury, the Confederates had captured his train, including his headquarters wagon which contained private letters. Some of the spicier ones from Libbie were reprinted in a Richmond newspaper.

On the Washita, faced with unknown strength, and isolated from any assistance, Custer correctly decided to conclude his victory and remove his command from further risk. No one knew what had become of Elliott, but it was hoped he would turn up. Captain Myers had been ordered to search to the east and had gone nearly two miles, but there had been no signs of the major and the other missing men. In any case, the chance of springing a further surprise on the Indians had become impossible, and increased casualties would be the only result of remaining in the area, much less pressing farther east.

Once the 7th had been reassembled and the prisoners mounted, by now after dark, around 9:00, Custer nevertheless proceeded to march his command east, bugles blowing and flags flying, in the direction of the remaining villages.

The ploy worked perfectly, for the Indian warriors who had been harassing the regiment throughout the day suddenly disappeared.

They raced to get back to their villages and their families, to defend against what they feared would be another cavalry onslaught on their own camps. After about an hour, the 7th Cavalry reversed its direction and vacated the field, heading for a successful rendezvous with the wagons it had left behind on the plain.

On December 2, 1868, General Phil Sheridan and everyone else then resident at Camp Supply was witness to a triumphal procession that might have competed with any Julius Caesar had staged in ancient Rome. As the 7th Cavalry approached the camp on a mild, clear day, it was preceded by its Osage scouts, in full paint, dancing and circling their ponies, waving spears adorned with scalps. Next came the white scouts, in all their rugged, individualistic attire. These were followed by the band, playing "Garry Owen." Custer, dressed in buckskins, rode at the head of the regiment's 53 prisoners, and then the troopers themselves in columns of four, the sharpshooters in the lead. The wagons brought up the rear. Sheridan later wrote to Custer that "the Battle of the Washita River is the most complete and successful of all our Indian battles . . ."

Custer subsequently came under severe criticism from certain cavalry officers, notably Captain Benteen, for failing to rescue Elliott and his men (or at least to have recovered their bodies). Although the outcry from the East over the destruction of the Cheyenne village was also predictably fierce, the fact remains that the raid on the Washita was a huge success for the U.S. Army on the southern Plains.

By the end of Sheridan's winter campaign, the Commanche, Kiowa and Arapaho had settled down on the lands assigned them by the government. The Cheyenne raided Kansas again the following summer, but after another attack on one of their camps—by the 5th Cavalry at Summit Springs, Colorado in July—many of them came in to Camp Supply in the fall of 1869. Others went north to join the Sioux, who were the largest tribe remaining defiant of the whites.

As for George Armstrong Custer, he had now taken on a new persona. He was no longer the "boy general" who had performed so dashingly in the Civil War; he was now the "*coures de sabre* in fringed buckskins," and considered to be America's greatest Indian fighter.

He remained a center of public attention with additional colorful exploits, together with the 7th Cavalry, sandwiched around a two-year assignment at a post near Lexington, Kentucky, where he wrote his Western memoir. Still, Custer didn't establish himself as a household word for generations to come until he and his entire command of the 7th Cavalry were wiped out, primarily by Sioux led by Crazy Horse and Gall, at the Battle of the Little Bighorn. This was one of the momentous defeats in U.S. history, and the largest in which the entire army force was killed to a man.

When the 7th Cavalry found a huge camp of hostiles on June 25, 1876, in southern Montana, it was after midday and the chance of achieving a surprise attack was remote. Nevertheless, a tactical maneuver could still throw the enemy into confusion. At this point, it might be mentioned, Custer also had the option of doing nothing at all, except wait for infantry and other units to arrive, which they did two days later. But then he would not have been Custer.

After sending Captain Benteen with three troops to cover the left flank, he ordered Major Reno to attack the village head-on, pinning the Indians to his front. Then Custer with five troops, the largest contingent of the 7th, moved off to the right to get behind the enemy. It was his devastating attack against the Indian flank that would no doubt achieve victory, against however many warriors were present.

Reno's frontal assault, however, did not go well. He not only failed to pin the Indians down, they chased him back, first to a patch of timber and then across the Little Big Horn to a hill, where his men dug in for defense. Reno himself seemed to have become unnerved, perhaps at the moment when a scout he was talking to was shot in the head. (It was probably at this time, too, that Reno discarded his hat.) Custer, meanwhile, exploring his way along the right, had realized the enormity of the hostile camp, but was still determined to execute his flanking maneuver in support of the three troops he had already launched against the village. He did send an urgent message to Benteen: "Big village. Come quick. Bring packs . . ."

Benteen moved to the sound of the guns, but stopped when he got to Major Reno. The captain, who had been the harshest critic of Custer for not coming to the aid of Major Elliott at the Washita, assumed a leading role in the defense of "Reno Hill," while Custer, meantime, across nearly two miles of undulating ground, attempted to fight off the largest concentration of Plains Indian warriors in history.

Along with every other single aspect of the Little Bighorn battle, controversy and speculation attend the elusive facts about how and where Custer was killed, and what shape his body was in when General Terry's troops arrived on the scene. One must recognize the possibility that he was as gruesomely mutilated as any other soldier on the field; the champion in this respect, after all, was his brother Tom, who was recognized only through a tattoo. The belief persists that Rain in the Face fulfilled his grisly promise, made some years previous, to cut out Tom's heart, however this is unlikely. (The truth is that Tom Custer, the only Union soldier to have won two Medals of Honor in the Civil War, might have done any number of things in that battle to earn such a passionate revenge.) As for Custer, some believe that he was killed early in the fight, in an aborted attempt to cross the river to get to the village. And if his body had indeed been cut and battered, this news might have been suppressed out of consideration for Libbie, not to mention the American public.

The vast majority of accounts, however, maintain that George Custer was found on the hill where the "last stand" was made, together with many of his officers, and where the most signs of a cool-headed, coordinated defense were found. He had been shot once in the side of the breast and once in the temple (a bloodless wound, no powder burns). He hadn't been scalped and had only been slightly mutilated, with two awls stuck in his ears.

The best story, from the Cheyenne chronicler Kate Bighead, is that when the Indians were ransacking the battlefield—killing the wounded, stripping the dead, and mutilating nearly everyone—two women recognized Custer and stood by him, preventing other Indians from desecrating the corpse. They did stick the needles in his ears so that he would "hear better" in the afterlife. These were two southern Cheyenne women who had known him since the Washita.

8

Koos de la Rey in the Transvaal

BY JANIS CAKARS

War was inevitable in South Africa in 1899. Cape Colony Governor Alfred Milner had concluded a year earlier, ". . . there is no way out of the political troubles in South Africa except reform in the Transvaal or war. And at present chances of reform are worse than ever." England had locked horns with the Transvaal and Orange Free State over the issues of British suzerainty and political rights for her people living on Afrikaner soil, but at the heart of the matter was the question of imperial destiny. England wanted supremacy in South Africa. The Boers wanted independence. "The white tribe of Africa" had staked its claim and was ready to fight for it.

The Dutch who first settled near the southern tip of Africa in the 17th century had been joined by French Huguenots, as well as a smattering of Germans and others and developed into a distinct nation. The language they had spoken developed into Afrikaans. Given to claiming huge tracts of land for cattle grazing and farming, the Boers (the Dutch word for farmer) had started to spread from the coast to the interior. This migration was then spurred on further by Britain's takeover of the Cape during the Napoleonic wars and their subsequent reorganization of the colony. After 1836 the mass-migration was dubbed the "Great Trek." If the British administration seemed oppressive, however, frontier life was not any easier. They constantly clashed with Bantu tribes already living in the interior over land that was equally unforgiving. The frontier experience came to define how the Boers viewed themselves and instilled in them an independent, nationalist spirit.

The Afrikaners were skeptical of centralized, bureaucratic government, even under the Dutch. The English, however, brought institutional and cultural changes that were alarming to many Afrikaners and furthered their mistrust of foreign rule. They did not wish to see the policies of European governments thousands of miles away dictate the manner in which they lived. Their 1880 "Petition of Rights" read, ". . . liberty shall rise in Africa as the sun from the morning clouds, as liberty rose in the United States of North America. Then it will be, from Zambezi to Simon's Bay, Africa for the Afrikaner!" They were a distinct people and their republics were to be defended against anyone who might threaten them—Bantu or British.

Before 1880 the Afrikaners' military experience had come from wars with the surrounding Bantu. In these conflicts they developed the tactic of taking up an advantageous position in close proximity to the enemy so to provoke an attack against prepared positions. This method of keeping the initiative, yet relying on tactical defense, had been consistently successful, and it led them to behave similarly in 1899 at the onset of war with the British. Constant conflict with their Bantu neighbors also developed in the Boers valuable scouting and reconnoitering skills. An agrarian culture, wedded to the land, they already possessed a high degree of horsemanship, as well as marksmanship. What would prove even more valuable when the showdown with Britain took place, was that the wars with their neighbors had made the Afrikaners intimately familiar with the raid. As Christiaan de Wet put it, ". . . their sanguinary night attacks were not easily forgotten." This practice was primarily used by both sides for the capture and recapture of cattle, but at times escalated into all-out-war.

Another important innovation developed by the Afrikaners during their conflicts with the Bantu was the commando. The Boers maintained this organizational structure throughout the Bantu wars and to the end of their conflict with the British. These citizen-soldier units, led by elected commanders, created a standing army, experienced and ready for duty at a moments notice.

In 1877, Britain had annexed the two nearly bankrupt Boer republics—the Transvaal and Orange Free State—and the beleaguered new nations had been temporarily helpless to resist. In the British view, the financial problems of the Afrikaners gave weight to their claim that the Afrikaners were incapable of governing themselves. An increasingly nationalist Boer population disagreed. In 1880 they rose in defiance of the Empire, and the next year won a startlingly complete victory over British troops at Majuba Hill. The British, then under a liberal government, granted the Afrikaner republics a partly limited, but acceptable independence.

While not interested in pursuing a war in 1881, in 1887 British interest in the Boer republics made a gigantic new leap, when gold was discovered on the Witwatersrand. Thousands of *uitlanders* (foreigners, mostly English) moved to the Transvaal; mines opened and commerce boomed. About the discovery, Transvaal President Paul Kruger correctly observed, "Instead of rejoicing you would do better to weep, for this gold will cause our country to be soaked in blood."

The Boers were indeed wary of the huge British influx and the threat of domination it presented. The Transvaal allowed the foreigners in, understanding that their own lack of resources prevented them from exploiting the mines, but that they could still accrue the benefits from commerce and taxes they would bring. But the Boers denied political franchise to the newcomers in order to retain control of their own hard-won republic. The Orange Free State, though less affected in this manner, allied itself with the Transvaal, recognizing that, as small, isolated Afrikaner states, their fates were unavoidably intertwined.

To the British Empire—its loudest voices now rich entrepreneurs like Cecil Rhodes and Alfred De Beers—the very existence of the Boer republics had become a severe annoyance. They hindered the expansion of commerce and stability in the region; worse, they denied political rights to Englishmen.

Cecil Rhodes became Governor of the Cape Colony in 1890 and his pro-*uitlander* stance over the next few years aroused growing apprehension among the Boers. In 1896, in order to subvert the Transvaal, Rhodes joined a conspiracy with Dr. Leander Starr Jameson, who was to ride into the Transvaal with 500 men and stir up a revolt against the Boers, helped by the manpower of disaffected *uitlanders*. The raid, however, was botched from the start. President Kruger and all of Johannesburg had caught wind of it, and even Rhodes, at the last minute, had tried to call it back. The raiders rode straight into the waiting arms of hundreds of mounted, heavily armed Boers and were easily captured. This embarrassment to the Empire would begin the downfall of Cecil Rhodes, the man who dreamed of a British-held Africa from "the Cape to Cairo."

The Boers could savor their victory over Rhodes only briefly, however, as an imperialist of equal fervor, Alfred Milner, succeeded him as Governor of the Cape Colony and High Commissioner for South Africa. Milner, too, was an ardent supporter of the *uitlanders* and a vehement critic of the Boer republics. The existence of the republics ran contrary to his belief that the uncivilized world should be gathered under the umbrella of British rule; they, naturally, being the most capable of rulers. Autonomous republics within and adjacent to the empire were undesirable. (The British had by then surrounded the Boer states by annexing Zululand, Bechuanaland and Basutoland.)

However, Milner was to go even further and put a new twist on imperial doctrine. He believed that not only was it proper for the English to rule others, but that, conversely, it was improper for others to rule Englishmen, as in the case of the Boers and the *uitlanders*.

Milner's arguments were persuasive back in England, and made more so by the concurrence of Colonial Secretary Joseph Chamberlain. "Imperial Joe" Chamberlain had forged a ruling coalition with the Conservatives in Parliament and, upon taking the post of Colonial Secretary, cited his qualifications: "In the first place, I believe in the British race. I believe the British is the greatest of governing races that the world has ever seen . . . and I believe there are no limits to its future."

Due to their views of empire and their positions of influence, Milner and Chamberlain have been given much of the credit for the war that was to rage for nearly three years in South Africa, but this is not entirely the case. As with all wars, a combination of factors was at work. While Milner and Chamberlain certainly acted as catalysts, the Anglo-Boer War was fought, variously, for the *uitlander*, for the mine owner, for England's coffers and to avenge the humiliation of Majuba, but ultimately to promote the growth and stability of the British Empire.

On the Afrikaner side the motive for war is easier to discern. It was a war for their independence. However, the definition of independence and the degree to which it could be given away to avoid war was debated by the Boers. As the Volksraad (Boer parliament) tried to reach a consensus, State Attorney Jan Smuts met with Milner to discuss possible amendments to Afrikaner laws that would make *uit-lander* citizenship possible. These meetings however, accomplished little. A meeting between Milner and President Kruger produced even fewer results. While the Volksraad argued policy, Kruger began an arms build-up. His principal acquisition was 20,000 Mauser rifles from Germany. Like the British Lee-Metford, this weapon could hold and fire bullets in rapid succession, but it loaded with a clip, while the British model loaded bullets into its magazine individually. This, combined with better sights on the Mauser, gave the Boer riflemen an initial advantage over the British in the war to come.

It's ironic that one of the more vocal opponents of the war, Jacobus Herculaas ("Koos") de la Rey, would go on to become one of the Transvaal's greatest generals. A farmer with little education, but of keen wit and staunch character, de la Rey was a popular member of the Volksraad. From the rural western Transvaal, he and his constituents had little experience with *uitlanders*, so the dispute over them was of minor concern. If the question was whether to give them political rights or go to war, it seemed more prudent to give them rights rather than risk the nation's destruction. At one meeting, he and President Kruger engaged in a sharp exchange over the issue. Kruger accused de la Rey of cowardice, to which Koos replied that, if war should occur, he would be in the field fighting long after Kruger and his party had given up their cause. The argument illustrates both de la Rey's commonsense caution and his dedicated sense of duty—characteristics that would contribute to his success. Until war was inevitable, however, he remained opposed to it.

But war it would be. The British issued an ultimatum demanding that the Transvaal repeal all legislation affecting the rights of *uit-landers*, cease importing arms (through Mozambique), submit all further disputes to third parties for arbitration and give political rights to English residents of the Rand. Inexplicably, though, the British sent their demands through the mail via steamship rather than by telegraph. Before the message even reached South Africa, the Boers had issued their own ultimatum. They requested Britain to remove all troops from the border and send back its soldiers who had arrived since June 1. An answer was demanded by 5 o'clock, October 11. Five o'clock passed. Kruger announced, "It will be a fight that will stagger humanity."

In London, a member of Parliament stated, "I believe the war will be brief, we will be victorious and that such a result will be to the advantage of the Boers, the blacks and the British alike."

Koos de la Rey said, "I shall do my duty as the Raad decides."

The Boers hit first. While General Piet Cronje struck out across the veld to attack the city of Mafeking, he detached a body of men under Koos de la Rey to raid the armored train *Mosquito*, which was en route to the city bringing guns and ammunition for its defense. Koos intercepted the train, tore up the tracks in front of it and captured its supplies. The reluctant warrior had drawn first blood in the war that was to distinguish him as one of the most capable military leaders to fight for either side.

The Boers laid siege to three important cities, Ladysmith, Mafeking and Kimberly. Each of these cities was important as a base from which the British could invade the republics and as a point of transport, each situated on important railway lines. With almost their entire male population, from ages 16 to 60, mobilized and mounted, the Boers seized the early initiative. The reduction of the British into what would later be called "hedgehog" positions, however, disrupted their momentum.

Lieutenant General Lord Paul Sanford Methuen was sent to relieve Kimberly. Methuen had served in the Ashanti War of 1873–74, in the Egyptian campaign of 1882, on the Bechuanaland Expedition in 1884-85 and then in India. Methuen was the perfect cut of a Victorian soldier: tall, thin and a little dandy. A most polite and courteous fellow, he was popular with the officers with whom he served. As well, he had a tremendous sense of duty and was diligent, conscientious and dedicated. This was to be his first major command, and he excitedly sent a message to the defenders of Kimberly that they would be relieved in about a week.

Methuen was happy with the men given him, which included some of Britain's most distinguished regiments. His only complaint was their attire. The British had already learned that shiny buttons, buckles and insignia would only draw the deadly accurate fire of the Boer marksmen, and had taken to painting absolutely everything khaki. Methuen complained that he looked "like a second-class conductor in a khaki coat with no mark of rank on it and a Boer hat and in Norwegian slippers." Methuen, for all his positive qualities, was a by-the-book general who possessed the flaw of inflexibility, sneering even at the practical measure of painting his buttons.

Confidently he set out on his mission, following the railroad to Kimberly. He correctly deduced that there would be three points on the journey that would provide suitable ground for Boer attack. Actually, there were four, but due to Methuen's lack of imagination, and Koos de la Rey's abundance of the same, one would elude him.

On the evening of November 22, the British reached the first such point, Belmont. Boers under Jacobus Prinsloo had noticed that the enemy was following the railroad and Prinsloo was waiting for them. The two sides skirmished that night and went to bed preparing for battle the next day. It was recommended to Methuen that he attempt to outflank the enemy but he rejected the idea outright, saying, "I intend to put the fear of God into these people." He chose a frontal assault instead. The British were, in fact, victorious and beat the Boers back into the veld. But, it was later learned, they had by no means put "the fear of God" into them. The Afrikaners made a skillful retreat, aided by 800 men under Koos de la Rey who, seemingly out of nowhere, rushed in and successfully ambushed pursuing British lancers and mounted infantry. Methuen lost 75 officers and men killed, 220 wounded.

The fight-and-run approach was perfectly suited to the Boers' skills and would become their trademark, although General Prinsloo at the time considered the battle a defeat. The goal of the Boer commandos was to make the war too costly for the British to pursue. Their hope was that the British would consider the fight for the republics not worth their effort, just as in 1881. As long as the Boers could inflict more damage than they took, they were winning the war.

Continuing his march, Methuen stuck to the railroad for several reasons: it was a direct route, he did not have adequate maps of the area and he had been instructed to repair it as he went. The railway was crucial for transporting British supplies, and the Boers, recognizing this, had early on destroyed as much of it as they could. Two days later, the British troops came upon the second position suitable for Boer attack. As predictable as Methuen's route, the forces of Prinsloo and de la Rey were waiting atop a series of kopjes (hills) near Graspan.

Again the Boers were unable to halt the British advance, but this time inflicted even worse losses upon Methuen. Still trying to prove what the British army was made of, Methuen had elected to make another frontal assault, and his attacking troops suffered fifty-percent casualties. A British newspaper described the abandoned kopjes after the battle as "almost dripping in blood; not a boulder escaped its splash of crimson." The Boers lost only 21 men. Undaunted, however, Methuen pressed on.

The Boers, emboldened by their mild successes and not wishing to see Methuen any closer to Kimberly, now decided to make a stand at the next point where the terrain was natural for resistance, but less obviously so: the Modder River.

While Cronje, a hero of the first Anglo-Boer war, ostensibly commanded the defense, it was de la Rey's creative sense of tactics that would design, and then implement, the Boer plan. De la Rey had received no official military training, but he had a sharp mind and was keenly observant. Also, his was far from the "typical" appearance of a daring soldier. He was 52 years old, with a long, thick beard (starting to gray) and deep, kindly eyes. He was tall and conveyed an air of fatherly authority and wisdom. State Secretary Reitz said that de la Rey had been blessed with "the gift of simple speech."

This soft-spoken, pipe-smoking, reluctant warrior commanded respect that would serve him well with troops who, without pay and regular provisions, were expected to live on the veld for as long as it took to defeat the world's greatest empire. De la Rey was seen by his men, in his corduroy pants and broad-brim hat, as a humble Boer champion who happened to be able to lick the English. And his simplicity, accessibility and commonsense approach to battles were to result in unique combat leadership skills. His plans were both clever and prudent, and in the heat of battle his passion would rise, inspiring his men as he yelled, "God is on our side!" Tactically inventive and intensely committed to the cause, de la Rey also possessed the gift of simple courage. When rallying his troops from atop his famous pony Bokkie, no Boer soldier needed to doubt him when he shouted, "I fear God—and nothing else."

At the Modder River the Boers had assembled about 3,500 men, a force equal to approximately half of Methuen's. De la Rey's idea was that, instead of placing men and guns behind the river to defend the crossings, he put them behind the forward bank, using the river as a trench. The Boers under his supervision dug into the banks and converted surrounding farm houses and kraals (Bantu villages) into hidden artillery emplacements. The old warrior Cronje distrusted Koos and his untried initiatives, but made no attempt to alter his preparations. Methuen, for his part, expected the Boers to make their stand farther north, by Magersfontein, and as he neared the Modder River he felt completely safe. As he looked in front of him across the flat plain he saw nothing, the Boers remaining invisible under the bank of the river.

In the early morning of November 28, Cronje, to de la Rey's shock and chagrin, ordered two guns to be moved from their hidden positions. This gave the British their first sight of the Boers. The Royal Horse Artillery shelled the Boers moving the weapons, who then made a hasty retreat off and out of sight. Inconceivably, Methuen still expected nothing. He told Sir Henry Colvile, "They are not there." Colvile responded, "They are sitting uncommonly tight if they are, sir." As if scripted for Hollywood, at that moment the Boer line—four miles wide—opened fire. They had waited until the British came within 1,200 yards and then unleashed a tremendous hail of bullets from their Mausers.

Taken completely off guard, the British collapsed. Methuen ran around aimlessly, unsure of what orders to issue, while his men hit the ground, not daring to stand and face the lethal Boer fire. Methuen himself was shot in the thigh. Late in the afternoon, after an artillery bombardment and infantry charge, the British did manage to break through the Boer left flank and cross the river, but when night fell the action ended.

During the darkness, both sides planned for a resumption of the battle at first light. Methuen decided on a renewed attack but would not have the opportunity. General Cronje, who had missed the entire battle, off at a hotel in a neighboring town, was concerned about the Boer flank and ordered a retreat. This made de la Rey furious. He accused Cronje of avoiding his duties and not understanding the situation. The British had lost 70 officers and men, with 413 wounded, and at that point were pinned where they lay.

Of the scant Boer losses, however, there was one of grave importance to Koos de la Rey. His son had been mortally wounded. The next morning de la Rey telegraphed his wife, "Today there slipped to death so softly in my arms our beloved son Adaan

. . . How hard it still is for us all. But God has so decided."

The painful reality of war was setting in on both sides. But it had only just begun. Methuen pressed on to the last spot on the route to Kimberly that offered itself as a natural site for defense: Magersfontein. He had been joined by two additional battalions of Scottish troops, from the Black Watch and Seaforths, and believed that he would now finally be able to inflict a decisive defeat on the Afrikaners. But again Koos de la Rey would deny him.

De la Rey decided to dig a trench in front of the hill at Magersfontein and position his men there, at ground level, instead of on the high ground as would be expected. Cronje was opposed to the idea, but, fortunately for the Boers, Free State President Marthinus Steyn stepped in and sided with Koos. As de la Rey expected, Methuen's plan was to shell the hill, then launch his troops in another frontal attack. Early in the evening of December 11, the bombardment started and the Black Watch, followed by the Seaforths, the Argylls and the Highland Light Infantry, moved into position for an assault at dawn. As Methuen shelled the empty hilltop for two hours, the Boers sat safely in their trench. At dawn the British rushed the hill and had come within 700 yards when the hidden Afrikaner line suddenly opened fire. A member of the Highland Light Infantry recalled that, as men started to fall on every side, "Somebody shouted 'Retire!' and we did—well not retire, but stampede; 4,000 men like a flock of sheep running for dear life."

Methuen suffered 971 casualties, expended the greatest amount of artillery ammunition in a single battle of the entire war, and then found himself back the next day at the Modder River. Kimberly would not be relieved until February 15, and Methuen would never be placed in command of so many troops again.

Lt. General Lord Methuen, however, was not a man to give up, and was to spend the rest of the war trying to atone for the shame of Magersfontein. His diligence would keep him in the field long after most other British officers had served out their tour and returned home. He would command fewer men and be placed under the authority of officers younger than himself, but he would continue his service. And he would meet Koos de la Rey again.

By that time, 1902, the war was scarcely recognizable from the heady days of Boer victories in 1899. On January 10, 1900, Lords

Roberts and Kitchener had arrived at Capetown to take matters in hand. Thereafter, not only had the sieges of Ladysmith, Mafeking and Kimberly been relieved, but the British had seized every major Boer city, including the capitals of the republics. Cronje had surrendered his army on Majuba Day, February 27, 1900, and Joubert's army had been defeated and dispersed on the approaches to Pretoria. A frail and dying President Kruger had gone to Europe on a fruitless diplomatic crusade. Fulfilling the deepest pre-war fears of the Boers, the British Empire had mobilized enormous strength that was still pouring into South Africa.

But both sides, not just the Boers, had underestimated the other's will to see the conflict through to a successful conclusion. On March 17, 1900, Boer leaders convened in Kroonstad for a council of war. Here, Koos de la Rey and Christiaan de Wet successfully argued for giving up "conventional" warfare, including further attempts to maintain standing armies. The "old guard," led by Cronje and Joubert, were gone, as well as any Boers who had been faint of heart or physically unable to continue the struggle. The new field commanders would be de la Rey, de Wet, Jan Smuts, Louis Botha and other leaders of hardened, mobile commandos. And they would rely on the tactic that had given them success against the British from the very beginning: the raid.

Turning to guerrilla warfare was a practical, calculated move on the part of the Boers, conventional war against the Empire no longer being possible to pursue. Their goal was to maintain a bleeding sore on the Empire that could not be salved, until which time faraway London grew tired of its effort and losses. By restricting their limited manpower to raids, the Boers would need fewer men in the field, and would also present fewer opportunities for the British to inflict losses in return. Further, a raiding war would be nearly impossible for the British to terminate with a single decisive blow. The Boers also hoped that a third party might intervene on their behalf. In the meantime, the contest would become one of endurance.

On March 31, de Wet christened the guerrilla phase of the war with a raid on Sannah's Post, the water station for the now British-held capital of the Orange Free State, Bloemfontein. Soon after, de Wet raided and captured the town of Lindley, along with its Irish Yeomanry defenders. De la Rey performed a similar feat at Zilikat's

Nek, forcing the surrender of the Scots Greys. However, the raiders had no place to put their prisoners, or even means to feed them. The Boers generally took their weapons, horses, and sometimes clothing, and then let them go.

The British, having recovered from their initial setbacks to thoroughly defeat the main Boer armies, were at first appalled that their victories had not concluded the war. Nevertheless, if more steps were required, the British were willing to take them. After thousands of their men had already been lost, there was no question of turning back now.

If the British could not "hold" Boer territory, deep within the bleak territory of South Africa, they would at least render it useless for sustenance for the commandos. The Empire's scorched earth policy burned farms, crops, and slaughtered animals that might be used to help sustain the Boer fighters. The civilian population—women and children the Boer soldiers had left behind on their farms—was rounded up and placed in concentration camps. Not even the British Empire, which had poured 500,000 men into South Africa by 1902 (against a pre-war Boer population of 85,000) could occupy every single foot of ground. But they could, in fact, render it useless for their enemy.

The British also devoted their enormous resources to creating "blockhouse lines." (De Wet called them "blockhead" lines.) This network of concrete emplacements and barbed wire across the veld, designed to hem in roving Boer commandos and to protect the railways, met with mixed success. But it was still another challenge for the Boers to face, and another sign that the British Army now, irrevocably, controlled their country.

As Boer commandos continued to fight for their independence, British frustration led them to abandon the notion that the conflict should be a "white man's war." Long after the conflict's end, Kitchener (who was made commander-in-chief of British forces, succeeding Roberts, who had succeeded Buller) admitted to arming at least 10,000 Bantu men. The claim that only whites would be involved in the war for the Transvaal and Orange Free State had been incorrect from the beginning, since both sides used black Africans to drive wagons, dig trenches and engage in all manner of support duties. But the arming of the Bantu was another step.

Also employed by the British during the guerrilla phase of the war were Afrikaners who had surrendered and then enlisted in the National Scouts. De Wet remarked in his memoirs, "I cannot resist saying that the English only learnt the art of scouting during the later part of the war, when they made use of the Boer deserters—the 'Hands-uppers.' These deserters were our undoing."

Throughout 1901, Koos de la Rey still roamed free in the western Transvaal, a seemingly unstoppable marauder against increasing numbers of British columns sent into his homeland in pursuit. Here, he earned the nickname "The Lion of the West." With his large commando comprised of "bitter-enders," Koos continued to deny the British victory in his region, just as de Wet, Botha and Smuts (who had ensconced himself within the Cape Colony) eluded the British in theirs. For almost the entire last year of the war, the job of stopping de la Rey would be given to none other than General Lord Methuen.

Methuen, like his Boer nemesis, was not a man who gave up easily. Since his humiliation at Magersfontein he had devoted his career to atoning for his failure. The senior officers who had come with him to South Africa had since gone home, but Methuen stayed, determined to make a difference and see the crown through to triumph.

Koos de la Rey and his men were meanwhile living at the very edge of their abilities. Their resources were more scant than would be tolerated by a population in peacetime, much less fighting men expected to operate with strength and efficiency. One British soldier, commenting on the state of guerrillas captured early in 1902, said: "Their remaining costume was in the last stages of decay . . . They were emaciated and drawn with hunger and hardship . . . But what is more humiliating than anything else is the realization that these miserable creatures are an enemy able to keep the flower of England's army in check."

The Boers adapted. Many of their Mausers had long since worn out or been lost. They were now armed primarily with captured Lee-Metfords. Clothing was also scarce (De la Rey's wife had even made dresses for their daughters out of captured Union Jacks.) and, after raids on British convoys, soldiers would be sent back without their uniforms. Food scarcity was also constant problem.

At this stage, Koos de la Rey's tactics also changed. On his order, his men took to firing from the saddle, always within the context of a sudden, surprise attack. They would now not be called on to stand against the British, but only to hit them, quickly, and when they were vulnerable.

At the end of February 1902, de la Rey had been quiet for some weeks, attempting to conserve his ammunition and the strength of his men. He kept his commandos out of sight most of the time. Methuen continued his hunt of the Lion, burning farms and killing cattle along the way, but without encountering his nemesis. With his skilled network of heliograph-equipped scouts, de la Rey was able to remain aloof from the British columns that twisted through the Transvaal in his pursuit. At the same time, he managed to hit a supply column here; a separated detachment there, in order to fulfill his mission: keep his men alive and the cost to the British rising.

The British, meanwhile, had been devastating the Orange Free State (now renamed by them the Orange River Colony), largely in pursuit of Christiaan de Wet. The Free Staters called on de la Rey to relieve the British pressure. Koos responded and, on February 25, 1902, an opportunity presented itself.

General Methuen had been called to Vryburg for administrative work and had left his command outside Klerksdorp. In the early morning hours de la Rey, with 1,200 men, trailed a convoy as it left the British camp to gather supplies from the town. The convoy was empty, save three carts loaded with ammunition, and was escorted by 700 soldiers. Koos sent some of his men ahead with orders to dismount behind bushes at the roadside. When the convoy passed the scrub, these men unleashed their fire on the rear and center of the British column. Then Koos charged with the remainder of his men, firing from the saddle. The wagons collided and caused a pile-up of vehicles. Despite a valiant stand by men of the Royal Artillery, who stood their ground calmly in the midst of chaos, the raid was a complete success for Koos. The British suffered 178 dead or wounded and nearly 500 captured. All the British horses, mules and ammunition were confiscated by the Boers.

The next day, over the protests of some of his men, Koos let all his prisoners go. The Boers themselves, if captured, knew they would be sent to St. Helena, or worse. The chivalrous Koos de la Rey, however, would not tolerate revenge seekers and those who mistreated British prisoners were flogged.

When Methuen heard of the disaster he immediately set out after de la Rey. He departed Vryburg with 1,300 troops, six guns and eighty wagons. Troops formed a screen around the wagons (pulled by both oxen and mules) in the slow-moving column. The soldiers Methuen had available to him were mostly inexperienced and ill-trained; but he intended to join forces with Colonel Grenfell and another column of 1,500 men just south of Lichtenburg, de la Rey's hometown. They were to rendezvous on March 7. The hunt was on.

The Lion of the West, however, was fully aware of what was going on inside his den. As Boer scouts tracked Methuen's progress, 1,100 hand-picked men of the de la Rey commando mounted up. While the British were unaware of de la Rey's exact whereabouts, the Afrikaners already had the British in their sights.

On March 6, the Boers were in place between the two converging British forces, then still 25 miles apart. That night Methuen camped at Tweebosch. Not far off, but undetected, sat Koos de la Rey, pondering his next day's specific course of action. The Boers could attack either of the British columns, or they could simply avoid them both, continuing the cat-and-mouse chase that had frustrated the British for over a year. But for de la Rey, it was not a difficult decision—the Lion could be hunted, but was also, himself, a hunter. Methuen would be the target at dawn.

At 4:30 the next morning, Methuen broke camp and continued north. He had just approached a confluence of the Little Hart's and Great rivers when, with the first rays of sun, shots were heard at the rear of his column.

Koos launched his attack with skirmishers dashing out against Methuen's rearguard. The British column halted, confused. Then, just as the British were feeling their first sense of fluster, a line of Afrikaners charged out of the morning haze across the length of their right flank. The rapid fire from the blur of Boers flying past the column tattered the screen of troops defending the wagons and guns. Then suddenly, a heavy fire opened up on the left, where de la Rey had placed three successive skirmish lines, the true direction of his assault.

The colonials and the yeomanry within the column were already beginning to melt when a mass of Boer horsemen, firing from the saddle, came charging through their own skirmishers on the left flank. These mounted Boers overran the column and, in and among the troops, began the chaos that would end with its destruction. With men falling on every side, their animals uncontrollable, and hordes of murderous Afrikaners seemingly springing from the ground, panic spread like a wildfire up and down the length of the British column.

De la Rey had timed his attack well. Methuen, cornered by the rivers, was soon enveloped by the Boers. As the column disintegrated, Boers leaped from their horses and pounced upon terror-stricken troops. The colonial troops were already fleeing from the scene; the yeomanry seemed to disintegrate.

But not all the British fell prey to panic. In the center of the column, General Methuen rallied his best troops around his two remaining guns, and attempted to fend off the attack, a valiant resistance in the midst of carnage.

A corporal of the Northumberland Fusiliers, who stood amidst this last stand, later recalled: "It was a dreadful sight around the guns, just like a slaughter-house. . . . The last gunner, finding himself alone, was just leaving when he was shot through the head. Lord Methuen did not quit the guns until then. He came over to us and stood about fifteen yards from where I was. Five minutes later he got his first wound—in his right side—and then tried to mount his horse. I do not know what he was going to do, but his horse was shot in the leg and he had to get off. A few moments later he got his second wound—in the thigh. . . . His horse was shot dead immediately afterwards, falling on him and breaking his leg. The doctor went to him to dress his wounds, but, before he had half finished he was shot too . . ."

The officer commanding the artillery had chosen death over surrender; Methuen himself lay helpless until dusty Boers, their rifles warm, walked up to examine his bleeding, stoic frame. Some yeomanry and colonials had broken out of the battle and fled to the shelter of a nearby Bantu kraal. But the Boers smashed it in with the British artillery they had captured the week before.

The raid was over by 6:00 A.M. The Boers had killed or wounded 189 British officers and men, capturing 600 more, plus had acquired four more British guns. The real prize, however, was Methuen, the highest-ranking British officer "bagged" during the entire course of the war.

When Kitchener heard the news of Methuen's defeat and capture, he took to his bed for 36 hours. He refused to eat and confided to his aide-de-camp that his nerves had "gone to pieces."

Ian Hamilton wrote, "I can't tell exactly how folks felt about all this in England. At Pretoria I do know that it really almost seemed as if everything might crash back into chaos. In fact, the whole issue of the war seemed now to hinge on the Western Transvaal."

De la Rey's raid against the forces of an occupying power had been executed with unequaled precision and energy; it impressed on the British that the Boers still comprised a dangerous military force, willing and able to continue the war. That this was not actually the case only gives further credit to de la Rey's strategy. The raid could have bring results beyond its limited objective; and for a time Koos still believed that the war could be won.

Filled with a sense of triumph, de la Rey's men, with vengeance on their minds, called for Methuen to be shot. The British had shot Boer officers in the past, including Gideon Scheepers, so ill that he had to be placed in front of the firing squad in a chair. They pointed out that Methuen had burned de la Rey's own farm and that his son had been mortally wounded at the hands of the general's troops. However, revenge was not a legitimate concern to Koos de la Rey. In fact, while dead British soldiers were generally untended by Boers of the Transvaal, who refused to join British bodies with their sacred soil, Koos had set aside a special plot on his nearby farm specifically for any British who found their end in the vicinity. A deeply religious man, de la Rey treated the enemy with the respect he believed all Christian men deserved.

Methuen was sent, under a flag of truce, and with a doctor, to the nearest British hospital station. Koos also sent a message of sympathy to Methuen's wife, expressing concern for the seriousness of the general's wounds.

Although no one knew it at the time, the war was finally about to end. The man who had executed the first Boer victory had just engineered the last. Kitchener, on the day of de la Rey's triumphant raid, but almost certainly before he knew of it, had sent to the enemy copies of correspondence between the governments of Great Britain and the Netherlands. The Dutch had offered to mediate a settlement to end the war. The British had declined but stated, "The quickest and most satisfactory means of arranging a settlement would be direct communication between the Boer forces in South Africa and the Commander-in-Chief of His Majesty's Forces." Kitchener attached no comment of his own, but the Boers understood the offer.

The Boer leaders, including de la Rey, met with Kitchener on April 12, 1902, in Pretoria. Meetings continued for a month there, and then at Vereeniging. A settlement was hampered and delayed by divisions within both parties. Milner was at odds with Kitchener in his desire for unconditional surrender. Kitchener, who had at one point all but asked to be relieved of his command, simply wanted the conflict to end. The Transvaalers, as well, wanted the most expedient end to the war, but were blocked by the Free Staters, who had a valuable ally in de la Rey. On the heels of his success against Methuen, Koos saw no point in giving up. They spoke of fighting to the "bitter end," of God's desire to see the Boers through and, if necessary, surrendering in the current conflict, only to plan to fight another day.

The grim reality of the situation weighed heavily on the Boers nevertheless. The toll on their civilian population had been too severe. Their numbers were greatly diminished and their farmland was nearly completely destroyed. The Boer commandos could not be physically sustained much longer, and if they all went down fighting that would result, in any case, in leaving the republics in the hands of the British. The conference also informed the stronger Boer elements of the desperate condition of their compatriots. Deneys Reitz, who had fought with Jan Smuts in the Cape Colony, was startled at Vereeniging, when he witnessed the condition of the Boers of the (eastern) Transvaal: ". . . nothing could have proved more clearly how nearly the Boer cause was spent than these starving, ragged men, clad in skins or sacking, their bodies covered with sores, from lack of salt or food, and their appearance was a great shock to us, who came from the better-conditioned forces in the Cape."

Then, on May 14, while Koos was still arguing to continue the struggle, a British column led by Ian Hamilton inflicted a defeat on his commando, still in the western Transvaal. When the Boers next assembled de la Rey spoke: "There has been talk of fighting to the bitter end. But has the bitter end already come? You must remember that everything has been sacrificed—cattle, goods, money, wife and child. Our men are going about naked and our women have nothing but clothes of skins to wear. Is this not the bitter end?"

The "end" came two weeks later, on May 31, 1902; the surrender terms were signed in Pretoria.

Upon hearing the news of the signing, de la Rey's remaining six children wept. His wife, Nonnie, painfully exclaimed, "Why was all the bloodshed, the suffering? What was the purpose of it all?" This was a question that would plague Koos. He would never entirely accept the peace.

In 1914, South Africa entered the Great War, fighting alongside Britain. Internal debate over South African involvement transformed into a Boer revolt against the British, led by Christiaan de Wet. Koos de la Rey journeyed to Johannesburg (it is believed) to join the rebellion, but he never arrived. He was mistakenly shot by police chasing a band of criminals. The uprising was quickly put down by his former comrades-in-arms Smuts and Botha, who had risen to prominence in the new government, selling the idea of conciliation to the Boers.

But if de la Rey's former compatriots did not honor him after the war, the British did. A statue of de la Rey was erected at the British cemetery on his farm and a hospital built in his name—for which Lords Methuen and Kitchener were the first to contribute donations.

9

Lawrence of Arabia at Aqaba

BY SAMUEL A. SOUTHWORTH

As the fearsome slaughter fueled by thundering artillery and chattering machine guns continued in 1917 on the stalemated Western Front, a separate war was being fought in a distant land. Instead of the mud of Flanders, there was the endless sand of Palestine; instead of the steady rains of Belgium, there was the unrelenting sun of the Middle East; and rather than British and French soldiers marching on squelching duckboards through sodden trenches to their jumping-off point, there were bands of nomadic Arabs traveling swiftly on camels. The German-allied Ottoman Empire had long considered the wastes of desert its own, and while the British campaign to unsettle the region had met with some success, it had stalled in Palestine, due to warring tribal factions and British ignorance of how to marshal them. It was at this time that one of the great eccentrics of history stepped in to earn the affection and respect of the nomads—by his charisma and bags of gold—as he led them in a series of raids that ultimately led to Ottoman defeat and the beginnings of modern Arab nationhood.

The man was Thomas Edward Lawrence, whose Arabic was quirky, who cloaked himself in colorful enigmas, and whose celebrity and success in the desert were ultimately to drive him into self-imposed obscurity. It was his ascension to fame rather than his flight from it in later years, however, that revealed him as one of the most fascinating military leaders in modern history.

Lawrence was an archeologist; a Fellow at Oxford; a Lt.-Colonel in the service of British Intelligence; a corporal in the Royal Tank

Corps; an aircrewman in the RAF (who helped perfect the search-and-rescue crashboat); a self-taught explosives expert; a careful student of history and literature; a top-flight mechanic and engineer; a pilot; a motorcycle enthusiast whose favorite bike, and the one he was to die riding, was a gift from Mr. and Mrs. George Bernard Shaw; an intimate of the political, diplomatic and literary lights of his day; and a military historian—as well as one of the greatest raiders who ever led a surprise attack. He was equally at home reading Latin or blowing up trains, excavating Hittite tombs or leading a camel charge. And his most daring exploit was a five-hundred-mile ride that culminated in the surrender by the Turks of the Red Sea port of Aqaba.

While the Lawrence legend continues to be surrounded by mystery and exaggeration, this much we do know: he liked and understood the Arabs, and was able to turn them into an effective fighting force. By employing Bedouin tribes as allies, together with his instinctive grasp of raiding tactics, Lawrence was able to create a successful new front in the Great War, tying down many times his number of enemy troops. The Turks, for their part, were forced to defend hundreds of miles of rail line and scattered outposts, never knowing where the next strike would come.

Although they had been present in the region since before the Victorian era, the British had scant understanding or respect for either the Turks or the Arabs. As a result, their naval and military actions in the Middle East lurched and sputtered, and then a main effort drowned in blood at Gallipoli, where Turks littered the beaches with dead and wounded Australian, New Zealand and French troops in 1915. After catastrophic losses of nearly a quarter of a million men, the Allies withdrew from that Peninsula, abandoning, too, any further ambition to assault the Turkish capital of Constantinople.

Meanwhile, in Persia, a large British force attempting to march on Baghdad became trapped at Kut (in present-day Iraq) for five months, 1915–16, while relief armies dispatched from India failed to get through. The 10,000 survivors surrendered to the Turks on April 29, and Baghdad remained in Turkish hands until 1917.

Aside from vested British interest in the Suez Canal, much of the importance of the Middle East, then as now, lay in its abundance of immense oil fields. The entire region had long been a disputed pawn in the "Great Game" of espionage and empire-building that the French, Russians and British had engaged in for over a hundred years, and the growing need for oil made it one of the absolutely vital pieces to be controlled by the combatants in World War I.

The difficulty of operating in the Middle East was not only the obvious harshness of the terrain, but the vast cultural chasm between Westerners and the region's natives. Fortunately for the British, the only people who understood less about the Arabs than themselves were the Turks, followed by their militarily savvy but somewhat condescending advisors, the Germans. The Ottoman Empire's brutal and thoughtless treatment of indigenous peoples had already been demonstrated by their massacres of the Armenians in 1915. (While the numbers are still a source of dispute, it would seem that close to a million Armenians were either killed where they stood or marched into the desert to die.) In a cruel land under a merciless sun, the Turks had a special reputation. Their soldiers were disciplined, savage fighters and some of their junior officers brilliant; and their high command, if no less muddleheaded than those of most other nations during the Great War, was particularly renowned for ruthlessness.

The British Empire, nevertheless, remained unintimidated by the Ottomans, although after Gallipoli and the failed campaign in Persia they needed another solution.

It was at this point that a young lieutenant in the service of British Intelligence who was familiar with the land, the people and the history of the Middle East was called upon to see what it would take to inspire the Arabs to rise up and support the Allied cause. The answer was money, as well as the ability to dress, eat and ride like the natives—in effect, to "become" Arab. And T.E. Lawrence was the perfect man for the job.

Lawrence had a talent for exaggeration, and claimed he was the bastard son of a nobleman, although this seems doubtful. He did have a demonstrable knowledge of archeology, castles and world history, having roamed England, Wales and France as a youth. He had also succeeded in getting a First in History at Oxford by dint of a thesis on the Crusader castles of Syria, the majority of which he had visited himself, on foot.

Lawrence had been in the Middle East since 1909, first on his own, and then as an assistant at various archeological digs. He soon developed a reputation for knowledge and appreciation of the Arabs employed at the digs, and set himself to learning their language. The servants teaching him quickly became too timid to correct him, and as a result it became easier for them to learn his version of Arabic than to teach him the proper way to speak the language. But he was a wholehearted supporter of the Arabs, and an indefatigable observer of their culture.

At the beginning of the Great War, Lawrence had been tasked with filling in the blank spots on the maps of Sinai, which he did partly by making things up. He was also charged with putting together a handbook on the Turkish Army, a task he approached more responsibly by interviewing prisoners, gaining valuable insight into the foe he was soon to bedevil.

The key to galvanizing the Arabs into a cohesive force was gaining the attention and help of one Ali ibn Hussein, whose family laid claim to being the traditional rulers of the desert. While accepting bribes from the British, Hussein was hoping to win acceptance from the Ottoman Empire of his claims to leadership. Instead, the Turks sought a more compliant leader, thus making Hussein an unwilling part of the Allied war effort.

On June 5, 1916, Hussein announced the Arab Revolt, throwing his 50,000 men against the city of Jeddah, which they took despite appalling losses, and Medina, which they failed to take despite fanatic attacks. Assaulting fortified positions was not well-suited to these desert nomads, whose previous tactics had consisted more of piecemeal riding, shooting and then running away. They were forced to wait in the south, where they had little effect on the British campaign to take Palestine, and contented themselves with hit-and-run raids on railway lines. It was clear, however, and particularly to the young lieutenant in British intelligence, that if the fierceness and mobility of the Arabs could be properly employed they could make a meaningful contribution to the war.

Aqaba, located at the junction of the Sinai Desert and Arabia, had already been identified as a port which, if taken, could create a direct route for supply and control of the Arab tribes. But while British warships stood off the Red Sea port and shelled it, an amphibious assault seemed imprudent in the face of its vast trench works. No doubt the experience of Gallipoli was in the back of the minds of the officers who were considering how to make the attack.

But Aqaba *could* be taken from the rear, if there was any way to cross the harsh desert and appear suddenly behind the defenders. Lawrence had been chafing under the burden of working as a headquarters intelligence officer, and he began to make himself as obnoxious as possible, correcting the grammar of his senior officers and ridiculing their knowledge of the Middle East. Meanwhile he formulated in his mind how best to exploit the situation—both for the benefit of the British and Arabs, as well as for himself. With the Turks spread out over the landscape, trying to hold their position in strongholds and along rail lines, and with a force of Arabs who were energetic and had the backing of the local population, it seemed to Lawrence that a mobile camel force could set out from Al Wajh on the Red Sea, passing inland through the desolate wastes, and then head west to approach Aqaba from the direction where it was least protected. In October 1916 he was transferred to the Arab Bureau, which was under the command of the British Foreign Office, and was dispatched to meet with the leaders of the Arab Revolt.

Lawrence found them to be a fascinating, charismatic and undisciplined lot. The Sherif and Emir of Mecca was Ali ibn Hussein, an effete and indirect man who seemed a little lost, but considered himself the ruler of the Arab people. This was a trace unrealistic. Of his sons, the third (Feisal) seemed the most impressive, and it was on the occasion of their first meeting that Lawrence began manipulating him. "How do you like it here [in Wadi Safra]?" asked the Arab leader. "Well," replied Lawrence, "but it is far from Damascus" (the site of the strongest Turkish base). An awkward silence ensued, there in the dark tent where they sat on carpets surrounded by Feisal's followers. Was this short, young Britisher taunting the great Feisal? What rank impertinence! Time and again, Lawrence was able to motivate the Arabs by pointed (and no doubt ungrammatical) exhortations questioning their manhood—and it worked.

Listening to tribal leaders from throughout Arabia and Palestine talking at Feisal's dinner table, Lawrence was able to conceal his identity while finding out where they stood. The long and confusing evening convinced him that they were all pulling in different directions, and that they were no fonder of the British than they were of the Russians, Turks or French (who had designs on Syria).

Eight thousand men served under Feisal, of whom only 800 were mounted on camels, the rest being tribesmen from the hills. All of them had difficulty obeying anyone but their tribal leaders. They might be induced to act together, but there was no guarantee that they wouldn't settle old scores in the heat of battle, or resort to looting instead of following up on an attack. But the dream of Arab independence was a rallying call to all of them, and it was this card that Lawrence played. After being taunted by the Arabs that he was a Turkish deserter—because of his khaki uniform—Lawrence adopted an Arabic headdress, the beginning of his near-assimilation into their way of life. This earned Lawrence the enmity of the British captain who transported him back to give his report on the Arab Revolt. It was hardly the first time his superiors had found the diminutive officer to be out of line. But eventually it would be the Turks who found him most objectionable.

Having given his report to higher authority, Lawrence was sent back to become Feisal's military advisor. Assisting him was a Captain Garland of the Royal Engineers, who had an alarmingly casual way with explosives, but was able to school the Arabs in their use. For the most part, Lawrence was self-taught, as evidenced by his method of getting tightly packed blasting powder out of its crates by setting off a small charge in one corner of the crate. Garland had a rough-and-ready way with both blasting and teaching, and despite poor health he was able to destroy the first Turkish train of the campaign. He died of a heart attack soon after. But Lawrence carried on his work, and developed a gift for demolitions that stood him in good stead on the many occasions he chose the correct spot for placement to send a Turkish train crashing into a ravine. In January 1917 he set his first charge and derailed a train (although by standing slightly too close to the blast he was forced to weather a rain of debris). Soon there was a price on his head. All the trains were guarded by heavily armed Turkish troops and, despite the *Boys' Own Paper* aura that surrounds some of his exploits, Lawrence was pursuing a deadly profession.

Now Feisal, with Lawrence's help, began to organize the tribes along the lines of a modern army. Although this flew in the face of centuries of Arabic fighting practices, the prospect of independence was a powerful incentive for the nomads to accept authority and act in concert with the larger war effort. Lawrence organized and led increasingly formidable raids on the Turkish rail lines, spreading havoc and flame wherever he could. He learned to ride a camel at breakneck speed, and once even managed to mount a Lewis light machine gun in a bucket on his saddle.

Lawrence began to dress and eat just as the Arabs did, using long flowing robes to dissipate the sun's heat and provide warmth during the frigid desert nights. He found it was possible to drink and eat only every other day, and to confine travel to the cooler times of morning and evening, when the weaker sunshine allowed for faster movement.

Feisal was an extraordinary leader, and Lawrence learned much about patience from watching him deal with his men. Another great Arab leader was Auda, head of the Abu Tayi tribe. He was a tall man with a beard and no teeth; he had previously had false teeth supplied by the Turks, but when the Arab Revolt was declared he had removed the unpatriotic dentures with a hammer. Auda considered his life to be a legend-in-progress, and everyone he met a potential hero. He was robust and charismatic, and charmingly listened to any amount of advice, after which he would do just as he thought fit. On night marches, the booming voice of Auda would serve to keep the party together as he sang of great fights and old raids while the camels lurched along under the white moon. It was Auda who announced to Lawrence that all things were possible with money and dynamite. Between Auda and Feisal, the British had bought the fealty of two mighty desert warriors. But the challenge was to use them in a meaningful way.

As the winter turned to spring in 1917, Lawrence, suffering from the harsh conditions of organizing and leading multiple raids on remote rail lines, succumbed to exhaustion and took to his bed for ten days. It was during this time that he worked out the plan for the attack on Aqaba, and when he recovered he set it into motion. He pointedly failed to make clear his intentions to his British superiors, feeling that if it worked that would be enough notice; if they all died in the trackless sands, well, there was no dishonor in that.

So Lawrence set out with Auda's tribe from Wejh to cover the nearly 500 miles that separated them from the port of Aqaba. Early in the raid, a camel that belonged to man named Gasim turned up without its rider. Lawrence didn't particularly like Gasim, or consider him a good soldier, but he nevertheless turned around and rode back to find the man. He found Gasim wandering, almost blind from the sun, a few miles back. He had gotten off his camel to relieve himself and then fallen asleep, worn out by the forced march. Lawrence mounted the slacker behind him and caught up with the camel train. It was by such acts that the Arabs came to know Lawrence as not only "the man with the gold," but also as a noble and fierce warrior who cared for his allies.

The raiders made their way across the endless waste, each family comprising a small unit of its own. When the unwieldy party reached Nebk, they rested while Lawrence went off with a bodyguard to the north, traveling carefully and avoiding Turkish patrols, slipping into towns at night to gather information and allies. After he returned, a raiding party of five hundred men prepared to ride to the west to take Aqaba.

A good illustration of Lawrence's gift with the Arabs is the story of the night before they set out for their goal. Auda had been telling one of his long, loud and pointless stories, and when he was done Lawrence began to mimic him. Evidently parody was not well known among the men of the desert, but it soon became clear that this little Englishman was making fun of the great Auda. Howls of laughter swept the caravan, and none laughed harder than Auda himself.

That first day they made sixty miles. As they drank their coffee that night, Auda chided Lawrence, who had been telling about the powerful telescopes that allowed Western astronomers to see thousands of stars that the Arabs could not. "But behind our few stars we see God," Auda replied.

The next day they came upon three wells that had been recently blown up. Someone knew they were coming. They had been counting on this water, and without it could not continue their journey. Fortunately the Turks had failed to choke a fourth well, but it was not enough for five hundred camels. Lawrence went down one of the destroyed wells and opened it so they could all drink. He also determined to make a feint to the north, attacking a rail line at Deraa, 120 miles away. With a small party of 110 men, on fast camels, he was able to accomplish the mission, but only after many false starts and by dint of Lawrence's preventing his party from attacking every Turk they ran into.

When they rejoined the main group, word came that 400 Turkish cavalry were setting off from Deraa to search for them. But one of the Turks' guides was in the service of Feisal, so the enemy galloped about in circles and never came anywhere near the raiders.

But word also came that the seven wells at Jefer (en route to Aqaba) had been blown. Lawrence had no choice but to press on, hoping that some of them could be cleared. Indeed, by working in relays, the wells were opened. The next obstacle was a blockhouse, but when it was attacked, the Turks drove off the marauders with rifle fire, and then, thinking this a small tribal raid, went to the nearest Arab village and killed everyone there in reprisal. The nomads fell

T.E. Lawrence's raid on Aqaba, 1917.

on this party as it returned, and took no prisoners. The blockhouse was then easily overwhelmed. Lawrence then set to work wrecking a lengthy stretch of rail, using five-pound charges and short fuses. As a result, ten bridges fell to the man known as "El Orens, destroyer of railroads." The group then turned to attack the next blockhouse, at Aba el Lissan, only to find that it had been strengthened with a large force of Turks. Once again, if they could not force this point, the mission would be over. As a raiding party, Lawrence and the Arabs could afford to pick their targets, but as a large group in the hostile desert they needed the water that each of these places held.

At dawn they began sniping at the Turks. Modern Enfields (a gift from the British) vied with ancient muskets as a steady stream of bullets sought out Turkish targets. A small group was sent to cut the telephone and telegraph lines behind the Turks. The sun climbed high and it grew hotter and hotter. Small sorties would try to dislodge the Arabs, but were driven back.

Auda came up to Lawrence and said, "All talk and no fighting?" Lawrence goaded him back by saying Auda's men fired much and hit little. At that, the mighty Auda worked himself up into one of his patented battle rages, and sprang into the saddle. "Come along, if you want to watch the old man work!" he spat contemptuously over his shoulder. Lawrence had no choice but to mount up and follow. Soon the entire party was galloping downhill at a fast clip in an impromptu charge. No one ordered it, but that was the way things happened with the Arabs.

Lawrence's camel soon outpaced the others, and he found himself alone among the Turks, charging and firing his pistol. Suddenly his camel fell stone dead under him, throwing him quite a way through the air; landing hard, he waited to be run over by the camels following, or killed by the Turks. The rhythm of the gallop had reminded him of a poem, and he hummed it to himself as he waited to die. "For Lord I was free of all thy flowers, but I chose the world's sad roses." But the bulk of Lawrence's dead camel diverted the other onrushing animals, and the Turks were soon killed in an orgy of destruction by the maddened tribesmen. Aba el Lissan had fallen.

Lawrence later found that his camel had been stopped by a revolver shot to the back of the head, which he had evidently fired himself in the frenzy of the attack. (One can sympathize with the camel he was riding the time he mounted a machine gun!)

From the few prisoners taken, Lawrence learned that very little stood between him and the port of Aqaba. But the Arabs had heard that there was nothing defending Maan, a few miles to the north, and they clamored to go there and loot. It was only with much difficulty that they were restrained. That night, Lawrence provided some evidence of his state of mind as he moved among the dead, carefully rearranging some of them into more seemly poses. Finally Auda drew him away, and the march continued.

One final blockhouse was taken at the height of an eclipse that Lawrence noted in his pocket diary. Then the steadily growing Arab force arrayed themselves around the Turkish lines at Aqaba. (Other tribes had heard of the raiding party's great success, and had come to join in the victory and, they hoped, plunder.) The Turks sent out a call for reinforcements and said that if help did not come in two days, they would surrender. Lawrence sent in word that if they didn't give up at once, he couldn't answer for the actions of the tribesmen. The Turks took this to mean that a massacre was in the offing and ran up the white flag immediately in response.

So a quiet, scholarly little man had inflamed and united the tribes of Arabia, and led them in a wild dash across 500 miles of desert, destroying rails and blockhouses at every turn, to force the surrender of a vital port. Lawrence went—robed as a bedouin—to British headquarters in Cairo to inform them, after the fact, of what he had done. The legend of "Lawrence of Arabia" dates from that time. He was recommended for the V.C., and eventually awarded the D.S.O. by his own country, and the Croix de Guerre with palms by the French. His reaction to these honors is best summed up by the Greek slogan he put over the door of his beloved cottage "Clouds Hill" after the war, which translated as "Does Not Care."

The Arabs had done their bit, and continued to work in conjunction with the British as General Allenby pushed north. On October 1, T.E. Lawrence entered Damascus in a Rolls Royce armored car and attempted to set up Feisal as ruler of all Arabia. But when Lawrence went to the peace conference in Paris after the war to stand up for Arab independence, it soon became clear to him that he and Feisal, as well as all the Arabs, had been mere pawns in the "Great Game." The French were given Syria, and the British held onto Palestine, Trans-Jordan and Iraq, feeling this was adequate repayment for their war efforts. They tried to mollify Feisal by making him king of Iraq.

Lawrence worked at diplomacy with Winston Churchill for a few months, and began the writing of his masterpiece, *The Seven Pillars of Wisdom*. But he soon became a huge celebrity, largely through the efforts of British poet Robert Graves and American Lowell Thomas, who outstripped even Lawrence with his exaggerations about the desert war. Thomas gave sold-out lectures on "Lawrence of Arabia," some of which the subject himself attended in London.

Lawrence soon tired of living up to his reputation, and decided to enlist in the Royal Tank Corps under an assumed name. When the press found out, he was thrown out of the Corps, but soon joined the RAF as an aircrewman, and went on to develop the crashboats that the RAF used to recover downed pilots. He continued to maintain extensive correspondence with the leading figures of the day, until he died riding his beloved motorcycle. He was trying to avoid hitting two boys on bicycles who had been hidden by a dip in the road.

But Lawrence's legend never seemed to die. The reality and the illusion have become slightly mixed over time, but the essential image endures: the scholar who donned robes and rode a camel, who led the Arabs by bribery and courage, and who set the desert aflame with his relentless raiding. All things are possible with money and dynamite—and the force of a dynamic personality.

10

Gunther Prien at Scapa Flow

BY TIMOTHY KUTTA

The sleek conning tower of the German Type VIIB submarine broke the surface of the North Sea at 2331 hours on October 13, 1939. Lieutenant Commander Gunther Prien took position on the bridge of U-47 and scanned the horizon to the west. The jagged coasts of the British Orkney Islands were silhouetted against the bright moonlit sky in the distance. Just before midnight Prien drew within 600 yards of the Ross Ness lighthouse at the mouth of Holm Sound. He ordered a slight course change and pointed the prow of the U-boat toward the sound. In the half light of the moon, Prien was searching for one of the two narrow channels that led into the harbor. Special Operation "P," the German raid on the British Fleet anchorage of Scapa Flow, was underway.

The German U-boat fleet had proved to be devastatingly effective during the First World War. At the height of unrestricted submarine warfare, in the spring of 1917, a handful of U-boats sank 800 ships, totaling nearly 12 million tons of Allied shipping. Had the campaign continued at that pace, the Allies might well have lost control of the commercial sea-lanes and, indeed, the war. Antisubmarine technology saved the day, however, but just barely. The Allies recognized the destructive potential of the U-boats and they specifically outlawed them in the Versailles Peace Treaty which was signed in 1919. Indeed, the British were so fearful of the submarines that they tried to outlaw the class of ship entirely in the Naval Treaties of the 1930s. Most Allied commands saw the submarine as a useful tool for reconnaissance and scouting, however, and submarines continued in the service of the major navies.

The Germans were most intimately familiar with the combat potential of the submarine and, even though they were legally forbidden by treaty, they continued clandestine research and development. Establishing a naval design company in Holland, the German government kept abreast of current submarine technology and built submarines for foreign governments. Although the efforts of the company were quite small, they kept the German Navy current, and when Adolf Hitler and his Nazi Party came to power and repudiated the terms of the Versailles Treaty, the Navy was more than ready to rebuild its U-boat arm. Indeed, U-1, the first of the new U-boats, was launched on June 28, 1935, only 11 days after the Germans and British signed a treaty allowing the Germans to again legally construct submarines. The first of the new U-boats were small coastal submarines with a limited range that could only be used close to shore to protect Germany from attacks by enemy warships. Nonetheless, bigger boats were already on the drawing board.

The commander of the new U-boat arm was Fregattenkapitän (Junior Captain) Karl Doenitz. An experienced submariner from World War I, Doenitz had stayed on in the postwar German Navy commanding torpedo boats and the cruiser *Emden*. He believed that submarines, working in groups called "wolf packs," could locate and destroy entire enemy convoys and rob seafaring nations such as Britain and the United States of their ability to carry out maritime commerce. Doenitz's views were radical and in opposition to many of the top German commanders, but he had free rein in his new command and was able to pick the best and brightest captains and crews for his new boats.

When the war began, Germany's naval rearmament was still far behind what the naval staff considered necessary to battle the English and French navies. Few of the major German warships were ready to go to sea by September 1939. Germany had only five battleships against the Allies' 22, two heavy cruisers against the Allies' 22 and a mere six light cruisers to take on the Allies' light cruiser fleet of 61. The submarine fleet was not in much better shape. Doenitz estimated that he would need 300 U-boats to bring the enemy's merchant fleets to their knees. But at the start of the war, he had 57 boats, of which only 26 were the new type VII, which could be used for long-range ocean operations.

If the small number of boats were not problem enough, Hitler ordered that the antiquated "Prize Rules" of submarine warfare be observed. These rules required submarines engaging neutral cargo ships to surface, inspect the target ship to determine if there was, indeed, contraband aboard and, if so, to rescue the crew before sinking the ship. These rules were, of course, impractical. A submarine's best defensive capability was to remain unseen during an attack. Surfacing in front of a neutral ship ran the risk of being rammed or at least having its position radioed to the nearest enemy base or aircraft. Inspecting ships took time, and while the inspection was in progress the surfaced submarine was an easy target. Even if the ship was determined to be hostile, there was no room on the sub for the merchant ship's crew. However, despite complaints from Doenitz, the Führer wanted the "Prize Rules" followed and those were the orders issued to his submariners.

It was soon obvious that until these restrictions were eliminated the U-boats would never reach their full potential or destroy enough commercial shipping to disrupt the Allied war effort. Unable to carry on a mercantile war, Doenitz began to concentrate his efforts on finding and sinking the major combat ships of the enemy's fleet. This posed an even more difficult problem for the U-boats. Enemy battleships and cruisers were faster than the U-boats and were usually escorted by several destroyers. Unless the submarine was lucky enough to be in position directly in front of an enemy ship, there was little chance it could get in position to launch an attack. Frustrated by the restrictions placed on his command, Doenitz could do little but search for an effective way to strike back at the British fleet

Scapa Flow had been a British naval base for more than a hundred years. Located on the northern tip of Scotland in the Orkneys, the great anchorage runs 15 miles from north to south and 8 miles wide from east to west. It is a deep-water harbor of 20 fathoms in most places, and no less than 10 in its major berthing areas. It is surrounded by the main Orkney island in the north and several islands on the east and west. The three main entrances—through Hoy, Hoxa and Holm sounds—give the fleet access to the North Sea in the east and the Atlantic Ocean to the west.

The Orkneys are an isolated and desolate group of islands subject to the ravages of freezing storms, and are inhabited only by the heartiest of herders and fishermen. However, the islands' inhospitable locale enhances their value as a strategic base because few people means fewer problems with security. Further, the bad weather that often shrouds Scapa Flow in clouds and fog makes aerial observation or attack extremely difficult. In addition, the access to both the Atlantic and the North Sea put the elements of the British Fleet stationed there in a position to intercept any ships trying to move into or out of the North Atlantic. (This was especially important before the Germans won access to French ports.)

Scapa Flow was, in fact, capable of holding the entire British Home Fleet. During much of the First World War the major elements of the fleet had been stationed in Scapa Flow as part of the blockade against Germany. At the end of the war, most of the fleet moved to the more hospitable climate of Rosyth in Scotland. When World War II erupted, the British were again obliged to counter a German attempt to move warships into the North Atlantic and the fleet took up residence in Scapa Flow once more.

The Germans realized the importance of Scapa Flow even during the Great War and made two attempts to slip submarines into the British anchorage. In November 1914, U-18, commanded by Kapitänleutnant (Senior Lieutenant) Heinrich von Hennig tried to slip into Scapa Flow behind a commercial steamer. An alert patrol boat detected the intrusion, rammed the U-boat and it sank with all but one crewman aboard. In October 1918, UB-116, commanded by Oberleutnant zur See (Junior Lieutenant) Hans Joachim Emsmann tried to enter the harbor submerged. He was detected by British listening devices, however, ran afoul of an antisubmarine net and sank with all hands.

The defenses had been formidable in 1918 and Doenitz had little doubt that they had been improved since. The harbor was known to be heavily defended with patrol boats, mines, underwater listening devices, shore batteries and a series of defensive booms, nets and block ships that regulated the traffic in and out.

Still, the British ships themselves had to come and go, and the defenses had to be lowered each time a ship entered or left the harbor. If the defenses were lowered to allow a friendly ship into the harbor, another ship—or an aggressive submarine—might follow close behind. And perhaps, Doenitz thought, there were other weaknesses in the defenses.

He began a careful study of aerial reconnaissance photos taken during the first week of the war as well as prewar charts of the harbor. He also ordered U-16 into the waters near Scapa to study the harbor entrance. When Korvettenkapitän Wellner returned to Germany and delivered his report on September 11, he repeated the conventional wisdom that although the nets and booms protecting the harbor were substantial, they were opened often to allow vessels to leave and enter. An aggressive submarine commander might well slip into the harbor during one of these openings.

Although the theory made sense, it was unpredictable and difficult to implement. The submarine would have to loiter at the entrance of a heavily defended target until the nets were opened. During that time patrol boats might well detect the sub and destroy it before it ever had a chance to sneak into the harbor. There had to be a better way to penetrate the harbor.

Armed with the latest intelligence reports, Doenitz and his staff went to work to find it. The possibility of sneaking under the nets was studied. Cutting through the nets was explored and sailing around the nets was surveyed. Although all of the plans held possibilities, none was firmly accepted. By September 18, however, all of the plans were far enough along for Doenitz to approach his superior, Grossadmiral (Grand-Admiral) Erich Raeder for permission to proceed, one way or another, with the raid. Raeder was confident in the abilities of Doenitz and quickly gave his approval.

With official permission to carry out the raid, Doenitz asked for additional reconnaissance photos of the harbor. He needed the latest pictures of the defenses as well as the number and types of ships in the harbor. It would do no good to penetrate Scapa Flow if the British fleet wasn't there. The second set of reconnaissance photos arrived on September 26. They were clear and detailed, revealing that while the nets and block ships protecting the entrances to the harbor were impressive, there were several points between the block ships in the channels and the shore line which might allow a submarine, riding on the surface, to pass. There was, for example, a 500-foot-wide gap between the boom in Hoy Sound and the northern tip of Hoy Island. Another gap was found in Switha Sound between the boom and the island of Switha. There was another gap between the Flotta shore and the boom in Hoxa sound. The most promising gaps appeared in Holm Sound. There was a 200-foot gap west of the most western block ship in Kirk Channel, between the tiny island of Lamb Holm and the Mainland and a 400-foot one between the blockship and the western shore of Lamb Holm. Either of these breaks offered an aggressive U-boat commander an excellent opportunity to penetrate the harbor without having to worry about the position of the nets.

It was obvious that the breaches that had been located offered the best possibility of success, but obstacles continued to arise. The tide entering and leaving the harbor often reached speeds of seven or eight knots. A U-boat on the surface would have difficulty maneuvering in such fast running waters. If the mission was to succeed, Doenitz would need an experienced captain who possessed excellent skills as a seaman.

It might have taken several more weeks to bring all the diverse elements of the operation together. But the timing of the raid was accelerated when Hitler, who was not particularly enamored with naval operations, told an assembly of submarine officers on September 27, 1939, that the Luftwaffe would take care of the British fleet. This comment was even more demeaning in view of the fact that a U-boat sank the British aircraft carrier *Courageous* on September 17, and the Luftwaffe had so far done no substantial damage to the British fleet.

Stung by Hitler's comment and wanting to demonstrate the effectiveness of his boats, Doenitz decided to launch the raid at the earliest possible moment. He began an intense search for a captain to carry out his raid. After reviewing many possible candidates, he settled on Oberleutnant Gunther Prien, the commander of U-47. Prien was an excellent choice.

Gunther Prien had entered service in the German Merchant Marine at age 15. He served on a variety of commercial ships and at age 24 in 1932 he became a ship's master. In January 1933 he joined the Naval Reserve as an ordinary seaman and quickly rose through the ranks. By the start of the war, at age 31, he was in command of his own submarine, U-47. Throughout his career he had demonstrated a remarkable ability to remain calm under difficult situations. His long experience at sea gave him the ability to control his boat and his placid demeanor made him a popular skipper, much liked by his crew. Doenitz had seen Prien on an exercise and remembered his ability to calmly reason out a difficult problem. Prien had also shown aggressiveness and determination during the wargames. All of those virtues made him the ideal man for the job.

On Sunday, October 1, 1939, Prien was ordered to report to Admiral Doenitz aboard the submarine depot ship *Weichsel*. He did not know the nature of the meeting but suspected a big operation was in the making. Upon arriving in the Admiral's mess, Prien was greeted by Doenitz, his planning staff and Kapitänleutant Sobe, his flotilla commander. Doenitz's first order to Prien was to listen to the briefing but not to make a decision as to the feasibility of the project. Having said that, Doenitz turned the briefing over to his staff. Prien and the staff poured over the charts and photos of Scapa Flow for the next several hours. All of the entrances to the harbor were studied, known defenses identified and the potential for entry at each of the possible gaps in the defenses analyzed. After several hours of briefings and study, Prien was given all the charts and maps and told to return on Tuesday morning with his decision.

Prien had already made his decision. The operation suited his talents and temperament. The following Tuesday, October 2, he met with Doenitz and accepted the mission. He had decided he would enter Scapa Flow on the surface through one of the gaps in Holm Sound.

After a brief meeting, both men agreed that the night of October 13/14 would be the ideal time for the raid. The Northern Lights would be at their weakest and the new moon would allow the best tidal conditions. With the time fixed, Prien departed to prepare his boat for departure. Six days later, on October 8, U-47 cast off from its moorings and set out for Scapa Flow.

The trip across the North Sea was uneventful. Prien submerged during the day and ran on the surface during the evening to avoid detection. The crew, already aware that they were on some sort of special mission because of the small quantity of supplies loaded aboard before departure, were briefed on the mission on Thursday, October 12, as U-47 lay in 120 feet of water just off the British coast. He told his men that their targets would be aircraft carriers, battleships and cruisers—in that order. They would sink enemy ships until their torpedoes ran out and then make their escape in the ensuing chaos. The crew was served a hearty meal, after which they made ready for the attack. Torpedoes were checked and reloads were placed in close proximity to the tubes. Radio codes and other secret papers were put in a pile and prepared for rapid destruction. Explosive charges were also placed throughout the ship to be detonated if the ship was in danger of imminent capture. By the afternoon of October 13, all was in readiness and the crew was primed for action.

At 1915 hours U-47 broke the surface of the North Sea just west of Holm Sound and ordered a northerly course to close on Lamb Holm and the channels nearby. Fifty-two minutes later a merchant ship was spotted nearby and the boat submerged quickly to avoid detection. The crew, tense and nervous, waited beneath the black sea as the merchantman passed nearby. The sound man listened intently as the screws moved away from the submarine and finally disappeared in the distance. Prien then ordered the sub to the surface at 2331 hours and started his run in. A few hundred yards in the distance, he could make out the Rose Lighthouse on the southeastern tip of the Mainland surrounding Scapa Flow.

Once on the surface, the crew worked liked a well-oiled machine. Five men were on the bridge while belowdecks the other 43 men of the crew carried out their duties with practiced efficiency. Guenther Prien controlled the ship from his position on the bridge. Next to him stood Eins WO (First Officer) Engelbert Endrass, who was responsible for aiming and firing the torpedoes. The Assistant Boatswain Ernst Dziallas stood behind Endrass relaying orders and keeping an eye on the surrounding area to ensure they were on course. Amelung von Varendorff was the Watch Officer on the bridge while Seaman Gerhard Hansel was the sole lookout.

As U-47 glided into the Sound, Prien adjusted course, slowed speed and searched the surface of the water ahead for the telltale signs of a block ship. Then, before him he could see the jagged shape of the block ship in the water. Below decks Boatswain Wilhelm Spahr watched the depth finder as the boat approached the ship. Suddenly, the bottom came up. Spahr sounded the warning just as his assistant on the bridge realized that the surrounding land was not the Mainland. They were in the wrong location. Prien, too, realized his mistake and ordered the boat turned hard north. The 750-ton boat strained under the turn but righted itself before it ran aground. Prien breathed a sigh of relief then passed the word that all was in order. The boat headed north and sailed quietly into Holm Sound.

Now almost midnight, U-47 passed Lamb Holm on the port side. Prien ordered another turn, this to the northwest, and within minutes the line of block ships protecting the entrance to Kirk Sound came into sight. The men on the bridge tensed as the U-boat approached the block ships. The broken shapes of the *Thames* and *Soriano*, lying stern to stern, were evident just below the surface as the tide moving across them broke into clearly defined swirls and eddies. The wires and chains attaching the ships to one another were not visible but were obviously there. Each of the ships was also attached to the shore by thick steel cable and these were visible as they broke the water near the shore and then disappeared into the shadows of the Mainland.

Prien had now reached a critical moment. In a matter of seconds, in the half light of a northern night, he decided exactly how to get around the block ships. The reconnaissance photographs had shown their location but no one could be certain about exactly how to cross them until U-47 actually got to within a few yards of the ships. Prien decided that he would cross between their two sterns. He ordered a course change and pointed the nose of U-47 at a 136-foot-wide channel between the *Thames* and *Soriano*. As the submarine came on line, the tide caught it and made steering difficult. Prien ordered the medium dip cell flooded to put his vessel deeper in the water and give better control. U-47 settled a little deeper and steadied as it approached the channel.

Suddenly, the men on the bridge could make out the shape of an anchor chain stretching from the stern of the northernmost blockship down into the water toward the other sunken hulk. Prien could only guess where the chain went after it went underwater. He adjusted course, hoping the hull of his submarine wouldn't foul the chain. However, as the boat entered the narrow passage between the ships, the strong current took hold and pushed the 750-ton boat to starboard. Prien ordered corrections to the rudder but the sub would not respond. In an instant the crew heard the sickening sound of metal grating against metal as U-47 ground up against the anchor chain of the block ship.

Only a few seconds were left before the U-boat was dashed against the block ships. Prien ordered the port engine stopped while the starboard engine was ordered slow ahead. The rudder was thrown hard aport in an attempt to get the submarine off the cable and back into the middle of the channel. The maneuver should have been effective, but the swift current was too much for the boat and the stern stayed firmly grounded against the cable.

In desperation, Prien ordered the air tanks blown. The area around the boat suddenly came alive with the hiss of compressed air and an explosion of bubbles. The hull lifted out of the water and the submarine moved forward, back into the center of the channel. In a few more seconds, U-47 had cleared the narrow channel and slipped into the British Fleet anchorage of Scapa Flow.

Prien steered the boat in a westerly direction, heading for the main fleet anchorage between Hoy and Flotta islands. The crew relaxed a bit, even though they were close to the enemy town of St. Mary's; but then, the boat was illuminated by the headlights of a car passing along the shore road near the town. The crew froze, waiting for the first shouts of an alarm. The beams of the headlights slid along the U-boat's hull. The Germans on the bridge could see a sentry walking his post on the shore nearby. Then, as suddenly as they appeared, the lights disappeared as the car moved off into the distance. The sentry continued to walk his post and no alarms or sirens were heard. The sleek U-boat, unobserved by the enemy, continued on its course.

As the submarine continued across Scapa Flow, anxious crewmen searched the horizon for the silhouettes of enemy ships. Despite their obvious proximity to the anchorage they could not see any vessels. When they'd moved to within 3,000 yards of the fleet anchorage they still could not make out any shapes. Prien and the crew were shocked and surprised. They could not believe that they had penetrated the most heavily guarded harbor in the world on a day when the entire enemy fleet had sortied out to sea. Prien decided to seek out ships in other parts of the harbor. He turned back to the east, and when he could again make out the mainland swung onto a northerly course to search for ships anchored in the training area near Scapa Bay itself.

The Germans had been in the harbor for almost an hour when the lookout sighted the faint outline of two warships on the northeastern horizon. The most southern lay 3,000 yards distant and was clearly a battleship. The ship's tall superstructure, four large gun turrets and single funnel identified her as a 31,200-ton "R" class battleship. Unknown to Prien, he had the HMS *Royal Oak* in his sights. The more distant ship also appeared to be a battleship, although its identity could not be clearly discerned in the hazy half light of early morning. This was the seaplane tender HMS *Pegasus,* but Prien would not get close enough to clearly identify it as such.

Prien ordered U-47 into an attack position. Endrass plotted the firing solution, using the torpedo aiming device on the bridge. Targeting the bow of the *Pegasus,* he ordered a spread of three torpedoes readied on the target. The fourth torpedo in the forward tubes was kept, in the event they were discovered and had to shoot their way out of the harbor. Endrass locked in the range and bearing and the firing sequence began. When all was ready, he pushed the firing button, hopeful that his calculations would put at least one torpedo into each of the ships.

The submarine shuddered and rose slightly as three heavy 21-inch G7a torpedoes shot out of its tubes and sped towards their targets. Down below Boatswain Spahr tracked the time to target with his stopwatch. A few minutes ticked away; they waited for the sound of the explosions. The nearest ship should have already been hit. Something had gone wrong. Then, abruptly, at 0104 hrs, a geyser of water, clearly visible to the crew, erupted near the bow of the HMS *Royal Oak.* A hit! The crew had hit an enemy battleship, but only with one of the three torpedoes fired.

Despite the poor shooting, Prien still had one torpedo in his stern tube and he swung his ship about to bring the rear tube to bear once more on the enemy ships. Shortly after 0106 hours, U-47 launched another "eel," this time from the stern tube. Although Prien had shortened the range, this torpedo also missed, like two of the previous three.

With three of the four forward tubes and the stern tube now empty, Prien had a decision to make. The noise of the first explosion would most certainly have alerted the defenders. He could retire now, beating a hasty retreat the way he had come, or sit calmly in the middle of the enemy harbor for 20 minutes while the forward tubes were reloaded for another attack. The decision was not difficult for an aggressive commander like Prien. Two anchored enemy warships were still within range of his ship and he still had torpedoes. He ordered the tubes reloaded. The order was almost unnecessary. His men were already hard at work, putting fresh "eels" into the three empty forward tubes.

Amazingly, the defenders were not alerted by the first explosion, nor were they searching the harbor for enemy submarines. The torpedo had hit the *Royal Oak*'s anchor chain, doing little more than waking the crew and scraping the paint. Most British sailors believed the first explosion to have been some type of malfunction aboard ship. The vessel was inspected, but most of the officers attributed the blast to some piece of machinery gone to pieces or some combustible material igniting in a store room. Those who suspected enemy action were looking to the sky for enemy planes. Hearing no aircraft engines, even those doubters went back to sleep.

Just before 0115, the tubes were reloaded and Endrass took a final bearing on the closest enemy ship. Prien brought U-47 closer and Endrass prepared another spread of three torpedoes aimed only at the British battleship. When all was ready, he pushed the firing button and the boat jumped as the burst of compressed air sent each of the "eels" on their way. Down below, Spahr punched his stopwatch as each torpedo left the tubes. Again, the seconds ticked by slowly as everyone waited.

This time there was no error. Precisely on schedule, at 0116, the first of the three torpedoes slammed into the starboard side of the *Royal Oak*. The second and third torpedoes followed in rapid succession. Despite her thick side armor, the force of the explosions literally lifted the huge vessel out of the water, ripping out its bottom. Within ten minutes the ship capsized and sank, taking with her 786 officers and men out of her wartime compliment of 997.

Prien and the men in the conning tower saw the explosions, followed by sheets of flame, and knew the battleship was finished. However, they did not have time to celebrate. They were still deep in the enemy harbor, which would now be fully alert and looking for the enemy that had penetrated its defenses.

Ordering a quick course change, Prien sent U-47 back towards Kirk Channel. However, he did not wish to waste time picking his way between the two block ships and chose instead to exit the harbor by the smaller channel between the southern block ship and Lamb Holm. The water there was deeper and he could make better speed through the obstacles. In their rush to escape, the Germans narrowly missed a mole in a small anchorage just before the channel, but aside from that the exit went smoothly and, at 0215, U-47 entered the North Sea.

Prien headed for deep water and submerged. He still had a few torpedoes left and decided to lie in wait for any ship attempting to enter or leave Scapa Flow. No traffic was observed, and U-47 surfaced at dawn on Sunday, October 15, to begin its return journey. A British submarine chaser suddenly appeared, however, and drove the U-boat back under the waves. The British vessel spent the morning dropping 32 depth charges, but none of them found their mark. By dusk the submarine chaser gave up and returned to base. Prien surfaced soon after and resumed his course for home. On the voyage back the crew nicknamed their captain the "Bull of Scapa Flow," and painted a red outline of a snorting bull on the conning tower.

The raid had a tremendous impact on the British. Members of Parliament and civilians alike wanted to know how the Germans managed to penetrate the most heavily defended harbor in the world. Faith in the invincibility of the British Navy faltered. At a time when the Admiralty should have been plotting a strategy to defeat the Germans, they were busy answering questions and studying the defenses of their main harbor. Obviously, if the Germans had done it once, they could penetrate the harbor again. Until the defenses at Scapa Flow were improved, the Home Fleet's base was changed to Rosyth. This alone added several hours to the response time of any British fleet unit trying to intercept German ships using the northern route to break into the Atlantic.

However, despite the cost and discomfiture inflicted on the British, the raid could have been much worse. The main British Fleet was not in Scapa Flow on the evening of Prien's raid. As they were returning from a major sweep of the North Atlantic, the weather turned bad and the fleet had headed for the calmer anchorage of Rosyth that night. Had the weather been better, Prien might well have destroyed or damaged several major ships of the British Fleet.

Gunther Prien and his men returned to Wilhelmshaven on October 17 to a hero's welcome. Prien was awarded the Knight's Cross of the Iron Cross by Hitler himself. He was the first naval officer to receive such an exalted award and each of the U-47's crewmen was awarded the Iron Cross Second Class.

Germany had a new hero and he was a submariner. Doenitz was ecstatic because Prien and his crew had vindicated the admiral's faith in the power of the submarine. The raid had also given submarines new prestige in the eyes of the Führer and Doenitz's ideas would henceforth be considered more earnestly in Berlin. Within a few weeks, the "Prize Rules" were abolished and Germany turned once more to unrestricted submarine warfare, as proposed by Doenitz. Through his daring raid, the Bull of Scapa Flow had not only put the U-boat on center stage in Germany, but affected the entire future course of the battle for the Atlantic.

11

British Commandos at St. Nazaire

BY SAMUEL A. SOUTHWORTH

On the night of March 28, 1942, a curious flotilla of British ships made its way up the Loire estuary that led to the German-occupied port of St. Nazaire. The dark shape in the lead was a motor gunboat, followed by a destroyer, and the 17 small shadows trailing along like goslings were launches and patrol boats. Each was packed with grim men who had blackened their faces and were armed to the teeth. They were not there to invade, or to begin the taking back of the Continent. They came to blow up the dry dock of St. Nazaire by means of a daring raid.

The lone ship among the boats was HMS *Campbeltown*, an old, broken-down lend-lease destroyer with a top speed of 20 knots. Her best days had come and gone, but she was to enjoy a glorious final night—and an explosive death. Packed in her bows were five tons of TNT, set in concrete and equipped with a delay fuse. The plan was almost childish in its simplicity: just drive the *Campbeltown* at top speed through the dry dock gates, leap over the sides, inflict as much damage as possible and then escape in motor torpedo boats before the big bang. A half-bright hooligan could have thought it up.

To carry it off, however, would require the best efforts of the oldest and youngest branches of Britain's armed forces: the fabled Royal Navy, weaned on a history of resolute and indomitable courage, and the newborn Commandos, who took their name from the Boers and their spirit from Winston Churchill, who had declared his intention to "set Europe ablaze" with raids large and small along the coast. The golden age of raiding had been launched by the Germans with their

172

Brandenburgers and paratroopers, and then developed into a fine art by the Anglo-Americans. World War II featured more organized raids than any previous war, and of all those raiders, the British developed one of the most storied reputations for unparalleled valor and determination.

During the spring of 1942, most of the world feared that the war was leading to an era of brutal dictatorship, as territory under the flags of the swastika and "rising sun" continued to expand. In the Far East, British, American and Dutch forces had been destroyed, routed or taken prisoner wherever they attempted to stand against the Japanese Army. After a coordinated series of lightning attacks from Hawaii to Burma, Japan had added most of Southeast Asia to its conquests in China, and controlled thousands of square miles of ocean.

Western Europe was in the thrall of Hitler's legions, some of which had continued across the Mediterranean to North Africa; in Russia, Soviet armies fell back deep into the central Asian steppe. Fighter pilots of the RAF had managed to discourage a German invasion of the English coast, but the British war effort, as well as sustenance for its population, now depended on keeping its maritime supply routes open.

The Nazi submarine force was using "wolf-pack" tactics against vital British convoys with enormous success. The specter of starvation, for lack of food, and strangulation, for lack of equipment and raw materials, loomed large. While the Royal Navy struggled to guard the shipping lanes, meanwhile, the battleships and cruisers of the *Kriegsmarine* were the terror of the Atlantic.

The mighty battleship *Bismarck* had been pursued and sunk by Royal Navy torpedo bombers in May 1940, albeit after destroying the battleship *Hood* and badly damaging *Prince of Wales*. But *Bismarck's* sister ship, the *Tirpitz*, still exerted enormous influence from her berth in a Norwegian fjord. Although it put to sea only infrequently, the *Tirpitz*, by virtue of its heavy armament and massive size, cast a dark shadow across British naval strategy in the North Atlantic. The mere threat of *Tirpitz* suddenly appearing in the midst of a convoy was enough to send tremors of anticipation throughout the fleet. One Allied convoy, PQ-17, was nearly annihilated when the Admiralty thought the battleship had put to sea, and withdrew its escorts from the merchant ships in order to confront the behemoth. As it turned out, *Tirpitz* had only shifted her berth.

German naval planners, on the other hand, found their options increasing with each new conquest by the Wehrmacht. After it had operated strictly out of the North Sea throughout the Great War, the Third Reich had added to the territory available for use by the *Kriegsmarine* not only the coasts of Norway and the Low Countries but, by mid-1940, the northern coast of France. And the one port on the Atlantic coast that had a dry dock large enough to repair the *Tirpitz* was located at St. Nazaire, France. The huge facility had been built before the war to tend to the liner *Normandie*, and was the largest of its type. In addition, the Germans had turned the port into a haven for U-bats, with shelters nearly impervious to air attack. It was thus to St. Nazaire that British planners turned their attention in the summer of 1941.

"Would it be possible," they wondered, "to land a party of, say, 300 men, and destroy the docks, as well as damage the sub pens?" The advantage of a raiding party over repeated missions by Bomber Command was the ability to place (by hand and, in this case, destroyer) charges at exactly the most vulnerable points of the *Forme Ecluse* (as the dry dock was known). Even the most accurate aerial attack at the time could do no more than rain bombs on a general area. And destroying such things as the pump house, winding house and other precise targets from the air was by no means a given.

However, only a desperate, highly trained and motivated unit could hope to sail into the well-defended port and cause the exact damage required to render the dry dock useless to the *Tirpitz*. As for whether a raiding party could get out again after fulfilling its mission, the optimistic view was that an accompanying fleet of small craft could load the raiders off, in the teeth of German fire, and ferry them back to England. To many within the Admiralty, the entire idea seemed suicidal. And the obvious candidates for the mission were the Commandos.

The initial idea for the raid is credited to Lt. Commander G. Gonin of the Naval Intelligence Department. He had first proposed it at a meeting that discussed raid ideas on July 7, 1941. That such a reckless plan was not scorned at this point gives a good indication of the cool desperation of the British that year, as well as the cold-blooded determination of the planners. At the time, however, it was tabled for consideration and given the code name "Operation Chariot."

In February 1942, Lieutenant-Colonel A.C. Newman of No. 2 Commando was alerted to the mission, and asked to report to Combined Operations HQ in London to begin working up a plan. No. 2 Commando had not been previously employed as a unit, but members of the newly raised unconventional warfare unit had seen action in Norway during the defensive campaign of 1940, and the raid on Vaagso. Since that time, a succession of training exercises had tended to frustrate the specialist raiders. While negotiating obstacle courses and bobbing around in the waters off Britain until they were good and sick was fine training for what lay ahead, the type of men attracted to the Commandos in the first place were the sort who wished to take the battle to the enemy, at any and every opportunity. Among regular Army officers, there was a great deal of resistance to the use of small parties engaged in what was considered "improper" war, but the Commandos were about to come into their own.

The Commandos were created as a stopgap measure by the British after their main forces had been compelled to evacuate the continent of Europe at Dunkirk. Plans for sabotage and raiding had been studied since about 1938, but it was not until June 1940 that the idea of a raiding force actually became a reality, albeit of the "red-headed stepchild" variety.

11 Commando was the first unit raised, its designation chosen in the hope that the Germans would assume there were at least ten more such groups, when in fact there were none. They were selected from territorial troops, and on their first raid were issued 20 tommy guns (to be returned in proper condition after the action) out of the forty then available in all of England. The High Command was worried that such a weapon would encourage the waste of precious ammunition. The Commandos blackened their faces with paint from a London theatrical supplier, and amused one another on their way to the hostile shore by mimicking minstrel dialect. Initially under the command of Admiral Sir Roger Keyes, an old and blunt sea dog who had had a hand in the Zeebrugge raid of 1918, the Commandos raided Boulogne to no great effect; struck various targets in Norway, to somewhat greater acclaim, and pondered any amount of absurd proposals, none of which had been approved. But one action stuck out, where seven men and a flare gun destroyed 200,000 tons of oil at Harfleur, and this

seemed to bode well for the future. With the right sort of troops, it seemed great damage could be done with minimal effort.

The action at Boulogne was particulary music hall-ish, with poor navigation, inconclusive contact with German seaplanes and bicycle patrols, and the return of some of the raiders a trace tipsy from having consumed their rum ration. They were at first refused entry to their home port, and then arrested because of their disheveled appearance. But from this inauspicious beginning was to grow a force that would change the nature of modern warfare by harking back to some of man's most primitive tactics.

In March 1942, the Commandos were just over a year old operationally, and feeling their growing pains. Their legendary training camp at Achnacarry in Scotland (which was to award 25,000 coveted green berets during the course of the war, at a cost of 40 killed in training) had just opened, and a change of command held great promised for the future. When Lord Louis Mountbatten took over Combined Operations from Admiral Keyes, he brought to the endeavor a fresh sense of ambition and drive, as well as impeccable (and valuable) social connections.

For small-unit raiding actions involving naval transport, the Royal Marines would have seemed to be the logical choice, but they were overtaxed and lacking in manpower at this time. So initially in the British armed forces there were Army Commandos and the LRDG (Long Range Desert Group), then Army and Royal Marine Commandos later in the war. Today, the Royal Marines and SBS (Special Boat Squadron) carry on the tradition of the green beret. Their intelligent, aggressive and decisive contribution to the Falklands campaign of 1982 shows that they are the rightful keepers of the flame kindled all those years ago. The British Army has had to content itself with Paras and the SAS, red and tan berets respectively, for Special Forces.

Both at the time of their founding, and in later years, a fair amount of nonsense has been written and filmed concerning these special soldiers. The legend of the Commandos needs some clarification. They were not blood-thirsty thugs, or violent criminals, or "cafe gangsters." They were volunteers whose training was dangerous, and they were hardy, intelligent men who could do anything that the average infantryman could do, but a little better. They were trained to rely on one another, and to prize efficiency, initiative and audacity. The average commando was exceptional, and his commanders nothing short of brilliant.

They could be reckless, as when they offered to remove a dead tree at their training camp as a friendly gesture to their host, a local nobleman. In using explosives, it is important not to give in to the tendency to overload the amount of blasting material—in the mistaken belief that "If a little is good, a lot will be better." In the case of the dead tree, the commandos' good intentions resulted in the breaking of every window in their host's estate, as well as throwing all observers of the blast to the ground. A somewhat sheepish apology and a promise to make good was followed the next day in typical Commando style by the careful replacement of every broken pane. (The outcry from local florists, whose greenhouses had been denuded during the night by men with painted faces, was another matter entirely.)

Their training was always with live ammunition, which was expertly rained very close to the trainees (in about 40 cases, perhaps *too* close). It could also be used to punctuate instruction, as when the first group of U.S. Army Rangers took the course. When one Ranger, after two polite (if loud) exhortations to sit in the bottom of a small craft to improve its stability, still failed to move from his perch on the rail, he was promptly shot in the rump. Just a glancing shot, but an unmistakable signal to the Americans that this school would turn out men who heeded their commanders.

Such irregular training as rock climbing, landing from small boats on jagged, wave-swept coasts and the "death slide" (a means of crossing a river by sliding down a rope) contributed to turning these men into extraordinarily versatile soldiers. Their subsequent success in every clime and country where they fought is solid testimony to their thorough preparation and the personal pride each of them took in what their green berets represented.

The most severe disciplinary threat to most of the men was the dismal prospect of their being returned to their original regiment should they fail in training. That men would prefer training to cover ten miles in full kit uphill in 90 minutes, rather than go back to their parent units, is most informative of the Commandos' esprit de corps. They respected teamwork and initiative, and were tolerant of eccentricity, if it was tactically useful. Men like "Mad Jack" Churchill, who used a bow and arrow to good effect in France, and Lord Lovat, who raised his Commando unit from wily Scots gamekeepers on his estates, and who always kept his piper by his side, became legends.

For the planning of the assault on St. Nazaire, Lt. Colonel Newman of No. 2 Commando was fortunate to be paired with Commander R.E.D. Ryder of the Royal Navy because the two men quickly took a shine to one another and worked together most efficiently. It soon transpired that because of the tides and phase of the moon, the attack would have to take place in late March; as a result, the planners switched to top speed.

The Loire estuary was notoriously shallow, so the draft of the vessels to be used was an important consideration. As a result, to accompany the *Campbeltown*, 16 motor launches, one motor gun boat and a fast torpedo boat were chosen for their shallow drafts, high speed and heavy firepower. The motor launches were armed with 20mm Oerlikon cannon and heavy machine guns. The idea was to sweep along the mudflats at high tide in the dark, thus avoiding the coverage of the numerous German guns trained on the regular channel. There was concern that the tide wouldn't be high enough, that the *Campbeltown* would bounce off the gates instead of penetrating, and that the explosive charges would detonate on impact, killing everyone for a considerable distance around.

Weighing the contingencies with no small amount of nerve, the planning accelerated throughout the first weeks of March. A demolition party of 80 men was chosen from among Commando Nos. 1–6, 9 and 12, and a covering force of 200 men from No. 2 Commando was selected. In addition, a small number of French nationals were to accompany the raid, and two correspondents were picked to record the adventure.

The port of Falmouth, in southwest England, was chosen for launching the raid, while far to the north, in Scotland, the demolition parties practiced their individual tasks. Some of the men would have to carry 90 pounds of explosive—a daunting notion in view of the chance of drowning or being shot. On a more positive note, it was decided to manufacture special magazine holders to accommodate the amount of extra clips needed to supply the Bren light machine guns and Thompson submachine guns. If there was to be shooting, the Commandos intended to hold up their side of the fight.

Lord Mountbatten was closely involved with the planning, and a measure of the importance he attached to the raid can be gauged by his final remark to Lt. Colonel Newman as the latter left London for Falmouth: "This is not an ordinary raid, it is an operation of war."

In Falmouth, the raiding party completed its training, which included scaling walls, memorizing routes and moving in the dark, as well as a boat trip to the Scilly Islands which made everyone bilious. There was also a mock attack against the docks at Devonport, during which everything had gone wrong and the "defenders" had repulsed the raiding party. This was viewed as a good omen. HMS *Cambleltown* did not participate, as it was deemed unwise to attempt the gate-crashing part of the attack more than once, when it would be done for real. Lt. Commander S.H. Beattie, who was to command the destroyer, contented himself with finding out how fast the old rust bucket could go. It must have been odd to take command of what was obviously an expendable ship, and prepare to do what captains are always trying *not* to do: ram it at top speed into the shore. But Beattie was more than equal to the task.

Finally, on March 26, the force put to sea for the voyage around the peninsula of Brittany, accompanied by two escort destroyers. Their cover story had been that they were a special anti-submarine task force. But once out of harbor, all pretense was dropped. They were spotted by an enemy sub, which was able to radio their course and position, but on they pressed. Fortunately, the Germans assumed that the task force was engaged in laying mines. A few French fishing trawlers were encountered, and their crews transferred unwillingly aboard the escort destroyers. On the evening of March 27, the force closed in on the coast of France, and, after loading the leaders of the raid aboard Motor Gun Boat 314, signalmen raised German flags as a measure of deception before leaving their escorts to await their return.

It was impossible to hide such a large flotilla, but by flashing messages in German to the effect that they were a German naval group with seriously injured seamen aboard, they were able to slip into the Loire River's mouth. A battery of coastal defense guns opened up on them, but once again the raiders were able to signal that they were being fired on by their own forces. The battery fell silent.

The value of this ruse was that it enabled the raiders to close with the port before receiving the full brunt of the shore batteries. The funnel stacks of the *Campbeltown* had also been cut to resemble a German patrol craft, and this, along with the unlikely nature of the raid itself, helped prevent a stiffer defense. The Germans were disinclined, in these pre-Dieppe days, to anticipate an attack on one of their heavily fortified ports, and were therefore taken quite by surprise. But of course one of the great values of the Commandos was the audaciousness of their plans, and it would have taken a mad psychic to foresee the raid on St. Nazaire. Be that as it may, shortly after 0120 the charade could no longer be upheld.

From flag lockers the white ensign of the Royal Navy was broken out, the German flags hauled down and the "white duster" streamed out for all to see. There could be no mistaking the intent of the little fleet now. They were only a mile and a half from the dock gates. At that point every gun ringing the harbor opened up, raining a hail of shot and shell on the flotilla.

Heavy anti-aircraft guns, pressed into a coastal artillery mode, splintered and set ablaze many of the little wooden boats as the lines of tracer careened across the surface of the harbor, converging with deadly effect on the attackers. The *Campbeltown* was largely impervious to these, but the small boats suffered grievously. Many sailors and Commandos died in the oil-slicked waters when their insubstantial craft came apart under the heavy fire. Nevertheless, aboard the *Campbeltown* the order was "Full speed ahead!"

As the aged destroyer groaned to answer the call of the engine room telegraph, the men aboard must have felt the singular thrill of anticipation that only ramming explosive-laden ships against metal dock gates can produce. Despite a torrent of enemy fire that gouged large holes in her sides and killed or wounded many of her crew and passengers, at 0134 the *Campbeltown* crashed the south gates at 20 knots—and came to a grinding halt. His TNT-packed vessel firmly embedded in its objective, Commander Beattie remarked, "Well, there we are. Four minutes late!"

Withour urging, Commandos leaped over the destroyer's bows and set to work on shore. Some laid charges to destroy the pump and winding houses, while others traded small-arms fire with the Germans. Terrific explosions signified that dry dock's shore facilities had been successfully demolished.

In the cloudy skies above, 64 RAF bombers were tasked with a diversionary raid, but due to the overcast they were unable to do more than drop a few bombs and fly off. They had been ordered to make bombing runs limited to one bomb per pass, but, with the danger of raining bombs on unseen targets, including the French civilian population, they elected to loiter as long as their orders dictated, but without the full diversionary effect that the plan called for. They were, indeed, unaware of the raid going on below them.

Meanwhile, the little launches were being hammered. Only five of the original complement remained, the survivors of a few, including Newman and Ryder, being able to land. Other boats had been shot out of the water by the shore batteries, or engaged by German gunboats. The launches had been divided into two columns, and the port side one suffered appalling casualties. Every boat but one in that line was smashed by the 88mm and 37mm guns that defended the dry dock. A point-blank duel between the little launches and the shore batteries only reinforced the old dictum that ships should not attack fortified positions.

But fortified positions can be effectively attacked by Commandos, and with the survivors of the initial approach now mostly ashore, a series of confused actions broke out around the dry dock. The stutter of machine guns mixed with the sound of explosions to indicate to the Germans, if there was still any doubt remaining, that the enemy had landed in unknown and violent numbers.

Six motor launches were given the task of securing a position called the Old Mole, a jetty to the west of the dry dock gates. But two defending pillboxes poured out such a volume of fire that only one of the boats was able to land. Captain W.H. Pritchard and Lieutenants Walton and Watson were able to get ashore with five men and sink two German ships before they were killed.

The headquarters party under Newman had disembarked from Motor Gun Boat 314 and rushed to occupy a building that Newman had selected from the map as his headquarters. Unfortunately, the Germans had also considered it the perfect place for *their* headquarters. Firing broke out as the outnumbered Commandos raced around the docks, setting charges, dodging from cover to cover and using up their ammunition at an alarming pace. Newman and his group were being fired on from a number of directions, including armed German sailors aboard two boats in the inner basin, as well as a number of rooftop machine gun emplacements.

The plan was working, insofar as they had rammed the dock gates with the *Campbeltown*, but the "wheels were coming off" for the Commandos on shore. Their ambitious selection of what turned out to be too many targets, combined with the devastating losses suffered by the launches, meant that the raiding party would be unable to defend themselves much longer, and would be stranded by the lack of surviving boats. As the odds began to mount, Lt. Robert Burtenshaw was killed while charging a group of Germans near the north caisson, firing his pistol while singing "There'll always be an England."

The race against the clock, as well as the battle against the enemy, has always been a task—and hazard—for raiders. The element of surprise has a short life, and can be turned around as the enemy recovers, and begins to show his own hand. And at St. Nazaire, the Germans were beginning to get organized.

Motor Launch 14, having been repelled from landing upstream, was searching for a new spot when a much larger German patrol boat loomed out of the night. Lieutenant I.B. Henderson, in command, deliberately drew alongside the enemy and the two craft fought almost muzzle-to-muzzle. Sergeant Durrant of No. 1 Commando, manning a twin-Lewis gun station, was shot almost to pieces, but refused to give up until he fell. Finally, with the entire crew dead or wounded, the captain killed and Motor Launch 14 about to sink, the survivors allowed themselves to be captured.

On shore, machine-gun and battery fire converged on the small groups with daunting accuracy. Troop Sergeant-Major Haines was able to use a two-inch mortar and a Bren gun to good effect in response, and gratifying explosions indicated to Newman that his demolition parties were getting on with their jobs. The winding house to the outer gate of the dock had been reduced to rubble, as well as other buildings nearby. Perhaps some of the enthusiasm for demolition charges that had caused embarrassment in the destruction of that tree near their training camp was now being put to good use. Lieutenant S.W. Chant and his party were able to blast the lock on the main pumping station and go down 40 feet of metal ladders where, despite having been shot through both knees, Chant was able to place explosives with such care that the building was entirely demolished.

It had by now become apparent that a return to England by sea was not an option, and so Colonel Newman ordered his surviving Commandos to break up into small parties to try to escape overland, via France and Spain. Characteristically, he closed this desperate briefing under fire by remarking, "It's a lovely moonlit night for it." He could have been discussing a high-spirited lark of a spring evening. Instead, he was suggesting that the survivors break through the town and try to gain the countryside, with every German for miles around hunting them, and then travel surreptitiously until they could reach a safe haven hundreds of miles away.

So they ran across the bridges leading to the town, avoiding contact if they could, bursting in and out of alleys and houses. Finally, a group of about twenty (including Newman) found themselves holed up in a basement tending their wounded. Newman's plan was to wait for the next night and then try to get clear. But he had already decided to surrender if cornered, out of consideration for the wounded. And before long, German voices at the head of the stairs indicated that the war was over for these men. They were put in trucks and taken to a central location in a café, where other Commandos were being assembled after capture.

Of the original flotilla, only six motor launches and Motor Gun Boat 314 (with Ryder aboard) made it back to sea, where they met the destroyer HMS *Atherstone*. Three of the motor launches went on to England without escort, where they joined their comrades brought back with the destroyer. It was a somber group that asembled in and around Falmouth, as both Commandos and sailors tried to sort out the details of their horrendous losses. However, for the enemy, the raid was not quite over.

Burning buildings and palls of smoke wreathed the port of St. Nazaire. Dead Germans and Commandos littered the dockyards as medical teams huddled around the badly wounded before carrying them away for treatment. Prisoners were interrogated as German soldiers methodically searched the town for any Commandos still hiding. And the destroyer *Campbeltown* was jammed firmly in the dry dock gates, the force of her ramming having lifted her bows out of the water.

The Germans could scarcely believe that the British had been so deluded as to attempt to destroy the largest dry dock in France simply by ramming a ship. And at what cost! German shore batteries had destroyed most of the attacking flotilla, and the marooned British landing party had been annihilated. Perhaps a few representatives of perfidious Albion were still scattering across the countryside, but their escape would hardly diminish the magnitude of the German victory against the ridiculous British raid.

But as the sun approached its zenith on March 29, the true purpose of the raid became clear. Forty German officers strolled on the decks of the destroyer, inspecting her, and 400 soldiers milled nearby, when the 10,000 pounds of explosives hidden in the bows of the *Campbeltown* went up with a gigantic roar. The cataclysmic blast from within the old Lend-Lease ship not only destroyed the St. Nazaire dry dock, but killed every German soldier within 200 yards. According to contemporary accounts, bits of flesh were draped over the town in a nauseating rain. The shocking explosion caused panic among the surviving Germans, who thought the Commandos had suddenly returned. The next afternoon a time-delay torpedo went off, followed by another an hour later. The Germans had by now become unhinged, and 300 French dockyard workers were shot in the panic that ensued.

As for the Commandos still on the Continent, five of the raiders were able to reach Spain, with the aid of French civilians, and eventually returned to England. The rest of the survivors sat out the war in POW camps, secure in the knowledge that they had done the impossible, and taken part in what has been called "the greatest raid of all."

But at a very high price. Thirty-four Royal Navy officers, 34 Commando officers, and 300 enlisted sailors and Commandos were killed, captured or missing, out of a force of just over six hundred— in exchange for which the port of St. Nazaire was all but destroyed. The mighty *Tirpitz* never broke out into the Atlantic, and was eventually disabled by British aircraft at one of her berths in Norway.

Five of the raiders received the highest British military decoration, the Victoria Cross, for their valor at St. Nazaire. Commander Ryder, Lt. Commander Beattie, and Able Seaman Savage, of the Royal Navy received the honor, as well as Colonel Newman of No. 2 Commando, and Sergeant Durrant of No. 1 Commando. Durrant's medal was awarded on the recommendation of a German officer who had witnessed his defense of Motor Launch 14.

There has been criticism of the daring but costly raid on St. Nazaire, focusing mostly on its concept. The execution speaks for itself. Some say that the relationship between the *Tirpitz* and the St. Nazaire dry dock existed more in the minds of the British Admiralty than in reality, and that there may have been no need for the raid at all. Others point out that too many tasks were assigned to the Commandos on shore, resulting in a dispersal of their strength and a decreased ability to defend themselves against German resistance. That resistance was, undeniably, underestimated.

But no amount of analysis can call into question the cool courage and gallant spirit that the British Commandos and Royal Navy demonstrated in March 1942. The raid astonished the Germans and electrified the Allies at a time when little else was going right with the war against the fascist powers. If Lt. Burtenshaw was correct, and there will always be an England, then the story of the remarkable attack on St. Nazaire will always be remembered as one of that country's most audacious exploits.

12

The Canadians at Dieppe

BY CHARLES WHITING

The first volunteers from Canada had arrived in the UK in late 1939. They had been raw and undertrained. Many of them had drilled back in their native country in their own civilian clothes. The few machine guns they possessed had been leftovers from World War I, as were their rifles. But they had been of good heart and eager to fight for "King and Country."

More and more had followed, still all volunteers. Again more training, usually in small-scale formations. The months passed into years and still not a sign of action, even as British, ANZACs and even Indians vied with the German Army in the Mediterranean and in North Africa. Naturally, being young men, the Canadians had gotten themselves into plenty of trouble. There had been brawls in bars; fights between them and British Tommies over local women; run-ins with the local gendarmerie. All overly predictable with bored young men. By the summer of 1942 there were nearly 200,000 "Canucks" in the tight, half-starved island. And still these volunteers who had come from halfway around the world to fight had not fired a shot in anger.

In that same summer, Winston Churchill, the man who would finally send them into action, was afflicted by one of his moods—"the black dog," as he called it. Up to now, the British Army had suffered defeat after defeat. It had been run out of France, Greece and Crete. In the Western Desert the fighting was going badly for the British 8th Army against Rommel, the "Desert Fox," and a few months previously there had been that tremendous blow to British arms in the Far East: the huge surrender to the Japanese at Singapore.

But Churchill had other worries in his own camp. On June 22, 1941, Hitler had invaded Russia. The Red Army had been pushed back hundreds of miles. However, to the surprise of most observers, despite losing millions of men and gigantic swathes of territory, the Soviet Union was still hanging on against the Wehrmacht. And Stalin wanted help. Following America's entrance into the war, the Russian leader appealed to President Roosevelt, who, with his military advisers, wanted to take the pressure off the hard-hit Russians by invading France: a "Second Front" at the earliest opportunity. It was to be the slogan of that terrible summer of 1942, and a lot of young Canadians were going to die on account of it.

Churchill was appalled by the Roosevelt suggestion. The British Army, with most of its resources in training, and fighting overseas, was hardly ready to tackle the cocky, victorious Germans in France. The U.S. Army was certainly not. Still, Churchill knew that he needed America's full-hearted backing in the war if England was to survive. He would have to something. But what?

Later, when it all fell apart and the great disaster of 1942 overtook the eager but green Canadian troops in the UK, no one would admit first suggesting the ambitious plan of invading a major German-held port on the coast of France in a quick "show of strength." Churchill, Alanbrook, the U.S. Head of the Home Force, Montgomery, Lord Mountbatten—all had played key roles in helping to shape the disaster to come. Yet not one of them ever stated unequivocally that he was the father of what was first called Operation Rutter: the attack by the 6,000-strong Canadian 2nd Infantry Division and two more experienced British commandos on the French port of Dieppe.

Working and planning in some sort of fool's paradise, the British and their Canadian comrades set about preparing for the great "surprise" attack in early summer 1942. It was all very haphazard. Many of the key Canadian infantry officers had to be fired because they were suddenly thought to be too old for combat. Weapons to be used in the attack were found full of grease, not having been cleaned for years. Briefly, an attempt was made by the Canadian commanders to train their troops in an attack larger than that of battalion strength, which was the largest formation they had used so far.

Montgomery, who was still in charge before he was posted so surprisingly to take over command of the British 8th Army in the Western Desert, didn't like the Canadian plan one bit. He felt right from the start that it was doomed to failure, although at the time he didn't express his doubts to his superiors in London. None of the top brass did—then or afterwards, when it all ended badly.

The days went by. The time of the assault grew ever closer. From their sealed-off locations on Britain's southern coast, the eager young Canadians started to embark on their craft. Not for long. Suddenly, out of the blue sky, zooming in at 350 mph, three Focke-Wolf fighter bombers—"tip and run" raiders, as they were called that year— swooped in to strafe the Canadian landing craft. One of them was hit badly. It burst into flame. Three Canadians were killed outright; others were wounded. Hurriedly, the brass made a snap decision. Perhaps the operation had been compromised. It was better to call if off—now.

It was. Montgomery, soon to leave for Egypt, breathed a sigh of relief and was heard to murmur, "Thank God for that." The Canadians, some disappointed, some relieved, were ordered to stand down. The first attack on Festung Europa had been called off. They were to return to their original stations immediately. They did so to be welcomed at their home ports by cheering sweethearts and girlfriends and old salts who had wished them well. The locals had obviously known about the "most secret" operation all along. Behind them, the Canadian infantry left their landing craft littered with the details and plans of Operation Rutter for all to see. Who cared? After all, the operation was off.

Not for long. The British, meanwhile, had surrendered Tobruk and its 33,000 defenders to Rommel's panzers on June 19. This was not Stalin's idea of a "Second Front" and the pressure from both Moscow and Washington was still on. In the U.S. capital there was alarming talk of the Americans making their major effort in the Pacific rather than against Germany, if the British didn't agree to do something in Europe soon. Indeed, Washington ordered that the first fifty British-trained U.S. Rangers would take part in the next attack. It was to be a symbolic gesture. In due course, the first American soldier to be killed in action in Europe, Lt. Edwin Loustalot of the Rangers, would die in the assault.

The attack was on again on orders from the British War Office. Again it would be Dieppe, an assault that some British planners thought was already compromised. Later it would seem they might well have been right—but that was later. Now the attack would be code-named Operation Jubilee. In essence, the plan was brutally simple. A frontal assault by Canadian infantry, supported by British Churchill tanks, up the beach and into Dieppe itself. The main assault would be accompanied by flank attacks by two experienced British commando units. It was risky. Even the most "gung-ho" of the Canadians and the planners knew that. Everything depending upon being able to surprise the German defenders.

On the night of August 18, 1942, the 250 small ships pushed off from the British coast. They sailed at speed, crossing seventy miles of sea. Surprise was to be their secret weapon. Dawn was to be the time of the assault. They'd spend perhaps five or six hours ashore, carrying out their various attacks, and by nightfall on August 19 they'd be back in their home ports, enjoying their triumph, telling tall tales to anyone prepared to listen and tucking into something special—bacon and eggs. That, at least, was the plan.

Tense and a little jumpy, the fleet passed through the German minefields. No problems. They were getting there. Then the little craft started to break into their attack formations: five waves in all. Still it was dark and the Jerries evidently suspected nothing. The British radar screens were blank. Naturally the men were keyed up. After all the years of training and waiting, they were going into battle at last. Some wrote letters, others checked their weapons for a last time and some prayed—but not many.

Then abruptly, startlingly, things began to go wrong. A small German convoy of vessels swept into the formation from the inky darkness. The Germans reacted first. A star shell exploded above the British craft, illuminating them in its harsh, glowing light. Hastily, the commander of the lead British gunboat rapped out an order. The German escorts were quicker off the mark. Their guns thundered. Scarlet flame stabbed the darkness. Within seven minutes, as the official report recorded, "S.G.B. [steam gunboat] 5 was a shambles." She couldn't steam more than six knots and half her crew lay sprawled dead and wounded in the smoking wreckage. And the Germans knew the enemy was on the way!

Now, as the attackers approached ever closer to land, the pre-dawn gloom was split apart dramatically, savagely. Searchlights criss-crossed the surface of the sea. Guns thundered. Tracer ripped in lethal red fury towards the landing craft like angry red hornets. All was confusion and chaos—and sudden death.

Still the Canadians pressed on. They were entering hell and they knew it, but they were undismayed. This is what they had been waiting for for so long. If a price were to be paid for the dread experience to come, they were prepared to pay it.

The Lovat Commando, almost a private army recruited by the Scots Laird from the workers on the great estates he possessed in Scotland, went in like gray predatory timber wolves. They rushed the shingle, still unspotted although the German fire had intensified to the center, where already scores of Canadians were being slaughtered. They started to scale the cliffs. Their mission was to find an exit and support the hard-pressed Canucks. They bumped into an enemy bunker. It was soon dealt with. A shower of grenades, and it was finished off in a matter of seconds. Another enemy pillbox was knocked out. The Lovat Commando was advancing swiftly. Nothing seemed able to stop them. With them they took half a dozen U.S. Rangers, two sergeants leading the way, and blasted through the roofs of enemy strongpoints.

On the other flank, the second British commando was doing well too. But as dawn came, gray and reluctant—as if some God on high hesitated to shed any light on the war-torn world below—trouble struck. Just as Ranger Corporal F. Koons accounted for the first German to be killed by an American soldier in World War II, radial-engined German FW-190s came roaring in at treetop height. Their prop wash lashed the ground below furiously. They spotted the commandos immediately and their 20mm cannon blasted away. Deadly white shells sprayed the Allied advance and the commandos began to take casualties. Brave and experienced as they were, they had difficulty in advancing against that wall of death. As khaki-clad figures crumpled to the turf everywhere, the unwounded began to go to ground. The steam was already beginning to go out of Operation Jubilee.

Now the Canadian infantry was coming in strength. Men of regiments with proud names, which had earned their battle honors in the bloody trench warfare of the "Old War," the South Saskatchewan, the Queen's Own Cameron Highlanders of Canada, the Essex Scottish . . . and the Germans were waiting for them!

As the landing craft of the Canadian Puits hit the shingle and ground to a stop, the entire horizon in front of them was obscured by the fog of war, split here and there by angry bursts of cherry-red flame;

the enemy snipers, all experts, popped out of their hiding places. They knew exactly what their targets were: Canadian officers and senior NCOs, easily recognized through their brass badges of rank. Officers and noncoms went down on all sides, as did the men. In a matter of minutes, most of the Puits' key officers had been killed or grievously wounded, writhing in their death throes in the blood-red pebbles, with the battalion reduced to the strength of two companies.

Undaunted, the virtually leaderless Puits pushed on into that murderous hail of fire. Some were without their trousers. They had taken them off for reasons of their own before they'd entered the sea. One man was wrapped solely in a blanket. Again they started to take casualties as the well-protected enemy attempted to prevent them from getting off the beach.

Before the attack, some of them had been told that Dieppe was held solely by second-rate troops, a combination of old German men and foreigners from the occupied countries now pressed into German service. But afterwards, as one complained bitterly, "I don't care if the prick firing at us was blind in one eye and minus one arm, he was still shooting at our guys from behind a wall of solid concrete!"

The official after-the-battle report said it in a more dignified manner, but even officialese couldn't disguise the courage of the hard-pressed, leaderless Puits. It read: "Notwithstanding that, all the men followed their leaders promptly, an assault regiment on the offensive was transformed in minutes to less than two companies on the defensive. It was impossible to move. They lay watching for signs from their leaders." But when those signs came, they moved.

One such leader was Lieutenant W. Wedd. He reached the seawall alone; all the rest of his men were dead and wounded. Direct fire was hissing towards him frighteningly. Undeterred, he prepared to carry out his task: silence the enemy pillbox that was holding up the survivors' attack. Standing upright—there was no other way—he flung a grenade right through the firing slot of the pillbox. There was a muffled crump, a mushroom of thick black smoke and a sudden silence. In the same moment that he was cut down by a last vicious enemy burst, the pillbox went dead. He had killed everyone inside it. Later they found his body crumpled like a bundle of wet, abandoned rags on the shingle, a mute testimony to his bravery.

For the Royal Regiment of Canada, which had also attempted to force that terrible beach of sudden death and destruction, everything went wrong from the very start. They had waded ashore "light-hearted and happy." Not for long. Private Harold Price, one of the survivors, remembered long afterwards, not the German fire, but the sheer size of the obstacles to be overcome before they could get at the unseen enemy. "I took one look and saw the wall and the steepness of the beach. It looked to me like a bloody mountain. We had never trained on anything like that. Nobody had ever mentioned anything about barbed wire and when we got there there was this concertina wire on top of the wall. That was some obstacle."

Still, like the Puits and the Camerons, the Royals did their best—and then some. The slaughter, as one survivor described it, was "incredible." A hundred men coming ashore from one boat into that rain of white death were killed or wounded within seconds. Casualties mounted dramatically and the Royals were getting nowhere. In the end the naval beachmaster thought the survivors had had enough, and ordered what was left to be evacuated. Ross Munro, one of the few who managed to get back to England, scrambled for a boat. "The hand of God must have been on that boat," he wrote later, ". . . the engines started up . . . and we slid back into deep water, as if it had been pulled by something out to sea." Afterwards he always thought it had been "the hand of God." But for many that grim morning God was obviously looking the other way. As Munro recalled: "There was an opening at the stern and through it I got my last look at the grimmest beach of the Dieppe raid. It was khaki-colored with the bodies of the boys from Central Ontario."

In the end the Royals lost 209 men killed, 100 wounded and 262 captured, twenty of whom died of wounds later in captivity. Today a monument stands where they fought and died. The inscription on it reads:

"On this beach, officers and men of the Royal Regiment of Canada died at dawn 19 August 1942, striving to reach the heights beyond. You who are alive on this beach, remember that these men died far from home, that others, here and everywhere, might freely enjoy life in God's mercy."

Next to the monument that dominates the beach at Dieppe is the shell-battered pillbox that took so many Canadian lives that terrible morning, and ruined the hopes of the Royal Regiment of Canada.

Field Marshal Gerd von Rundstedt, the German commander in the West, was usually a sanguine character. The author of many famous German victories over the past two years, he had kept his nose out of politics, preferring to use it to sniff the well-aged French cognac he delighted in. But despite his ancient wrinkled appearance and non-committal manner, he had a mind as sharp as a knife. And for once he was rattled by this bold Canadian frontal attack and the great losses the Canadians were taking.

To what purpose? Surely they were intent not solely on capturing the relatively unimportant port of Dieppe, which wasn't really suitable for deep-water vessels? Was the Dieppe fiasco a diversionary attack as a cover for the launching of the Second Front in northern France? Von Rundstedt had his agents everywhere. He knew the pressure that drunken "old sot" Churchill was under to do exactly this.

At six-thirty that morning he made a decision. His headquarters alerted two elite German panzer divisions for an all-out counterattack: the 10th Panzer and the 1st SS Panzer, the "Adolf Hitler Bodyguard" Division. Both had just returned from Russia, but both were more than a match for the Canadians, whose tanks had not even been able to clear the beach. Indeed, the Calgary men who manned the lumbering, underpowered and undergunned Churchills would lose forty of them before the day was out without the slightest success.

Von Rundstedt, as usual, had his ear to the ground. While the panzers prepared to move into the attack (they had no maps of the area, which made it difficult), he noted that the Luftwaffe had been sent into a complete flap by the sudden enemy assault and that the defending infantry was using too much ammunition. He wrote of one group that they "fired 1,300 rounds of shells in half a day's fighting." It was attention to such small detail that had given him his victories in the past.

But despite the shortcomings of his Dieppe garrison and the Luftwaffe—which was still virtually totally confused an hour after the assault had commenced (although the flyboys had received an early warning from the Kriegsmarine)—the aged German field marshal sensed things were going his way at the port. Indeed, although he didn't know it at the time, the Canadians were suffering so many casualties that they were becoming downhearted. More and more of them were beginning to surrender. As the Reuters correspondent covering the Calgary tankmen wrote: "The Calgary men . . . are getting quiet and not a little dispirited. Their pals are fighting and dying within their sight and they are just waiting . . . Some of them are even sprawled out on the canvas covers quite oblivious to the constant explosions of guns on the shore, guns at sea, guns in the air."

Brigadier General Churchill Mann of the Canadian 2nd Divisional staff was more explicit. Naturally he knew more of the big picture than the Reuters correspondent. He told U.S. General Lucian Truscott, who had helped to set up the Rangers and who would one day be an army commander in Europe, when the latter asked him how things were going, "General I am afraid this operation will go down as one of the great failures of history."

In one last desperate measure, the Canadian commanders decided to throw in their floating reserve: a Royal Marine commando, led by Colonel J.P. Phillipps. As one of his marines who lost his arm in the doomed attack recalled bitterly, many years later, it was "like throwing bloody good money after bad."

Colonel Phillipps recognized the hopelessness of the situation, but good marine that he was, he didn't protest. Instead he pulled on his white gloves which he always wore in action so his men would recognize their C.O. and leaped on the small deck of the first landing craft. The white gloves were duly recognized, but not only by the tense expectant commandos, each man wrapped in a cocoon of his own fears and apprehensions. The Germans spotted them, too. A hail of fire came winging his way and Phillipps, that brave but doomed man, realized that there was no hope for the last abortive attempt to land; the enemy was too numerous and too well dug in. He had only seconds to live. But the proud marine colonel resolved to halt his commandos and save as many lives as possible. White-gloved hands above his head, he started to signal for the attack force to turn about, the craft to make smoke.

For a handful of seconds, the noble figure remained upright, enemy slugs cutting the air lethally all around him. His men prayed he would survive. To no avail. He fell, mortally wounded, to the bloody, slippery deck. But even as he died, he must have known that he had saved at least 200 of his beloved marines from death.

Now the time had come to abandon that deadly wasteland, completely dominated by the fire of the triumphant German infantry who had come out of their holes and, running over the khaki-colored carpet of the Canadian dead, began to harry the hard-pressed survivors. Inch by inch, yard by yard, the survivors were forced to the shore. Some gave up and surrendered. Here and there a "bomb-happy" Canadian went crazy, seeking refuge from the carnage all around in madness. But only a few.

The rest fought, retreated, turned and fought again. But they knew there was little hope for them. They would be annihilated, if the boats didn't turn up soon and take them off. In a matter of minutes they would be at the water's edge, and thereafter would have no further room to retreat.

But their spirit, even in defeat, was still willing. The colonel in charge of what was left of the Essex Scottish, retreating like their comrades of the Camerons, the Puits, the Royals and all the rest, was having no ill-discipline in his regiment although he knew they were virtually lost. He ordered his men not to carry their rifles by the trail as they moved through the smoke and debris of that dread beach. Instead they would "march to attention" like Guardsmen, with their "rifles at the slope." Proudly! And they did, too.

It was now six hours since the attack force had landed, carried forward the assault, the first against *Festung Europa* since the British had been chased out of Dunkirk two years before. All the hopes and high aspirations of that moment, which now seemed of another age, had been dashed. The time had come to order a general retreat of what was left. The Canadians had been defeated. But they couldn't—wouldn't—be abandoned.

Waiting for the codeword "Vanquish" for the evacuation to be given, Lt. Colonel Hillsinger, an observer from the U.S. Army Air Corps, stood on the bridge of the British destroyer HMS *Berkeley*, as the craft took a direct hit on the bridge. The naval commander was killed outright, plus several of his staff. Hillsinger was hit. Suddenly he found himself lying on the debris-littered bridge, in some pain, but conscious and with his sense of humor, however morbid, still intact.

The big American stared down at his blood-stained trouser leg with a mixture of anger and consternation. His right foot was missing; then he saw it, complete with shoe, bobbing up and down on the wavelets a few yards away.

Two sailors came rushing up to his aid, as cries for help were being raised on all sides. The colonel didn't seem to notice. He continued to sit up and stare at the missing foot wearing the shoe. Suddenly, seized by a great anger, he pulled the shoe from his good foot and slung it into the water after the other.

The sailors looked down at him in bewilderment, as the colonel snorted: "Take the goddam pair!"

The American looked up at the two British tars and explained the reason for his angry gesture. "New shoes," he said. "Bought them this week. First time on. What d'yer know." All this time he seemed unaware of the intense pain he was probably suffering.

Naturally there were cowards among the 1,000 or so wounded, but most of the men were enduring pain like Hillsinger: stoically, even angrily. Besides, the lightly wounded could no longer expect too much attention. The ships' medics were running out of bandages and drugs. In England, hospital trains waited for the urgent cases, while local civilian hospitals were alerted for the expected flood of casualties. Soon the young, poorly trained VAD nurses and middle-aged, even elderly doctors, recalled from retirement for duty, would be working flat-out around the clock, trying to save as many young lives as possible. Some of the small destroyers were bringing in up to five hundred wounded, the men parked everywhere, even the boiler room, where there was space for a casualty to lie flat.

But the harassed Canadian staff officers were more concerned with getting away the able-bodied than the wounded. The former would live to fight another day; many of the wounded, due to their types of injury, wouldn't.

The RAF flung in fighter-bombers in desperation. Smoke screens were laid, volunteers struggled ashore at the end of the battle, trying to guide stragglers to the waiting boats, hidden by the thick gray choking smoke screen.

There weren't too many of them left now. Their numbers were decreasing by the minute. Death, exhaustion and surrender were taking their toll by the minute. Although the German report was not strictly true—there were still hundreds of Canadians in France, including the most daring and determined already making a dash for neutral Spain (for some of them it would take months to get there and incredible adventures)—von Rundstedt noted in his war diary: "19 August,

1740 hrs. No armed Englishmen remain on the continent."

Some time later the aged, incredibly wrinkled field marshal came to Dieppe personally. He looked at the dead, examined the wrecked British Churchill tanks (he didn't think much of them) and then stared out pensively at the debris-littered sea. No one knew exactly what he was looking for. Was he here to savor his triumph? Was he staring at England beyond the horizon, pondering what Churchill might do next? Uneasily his elegant, red-striped staff waited for him to speak. When he did so, his words were careful, measured, confident, but without a hint of gloating or triumph. He said, as if it were merely a statement of fact, "They won't be back now for a long time."

Standing at the fringe of that glittering entourage of staff officers and war correspondents, Hauptsturmbannführer Jochen Peiper of the 1st SS Panzer Division, "Liebstandarte," who chanced to be there in the company of his divisional commander, "Papa" Dietrich, just made out the remark.

As Field Marshal von Rundstedt raised his bejeweled marshal's baton, awarded him personally by the Führer, in some sort of salute, the young SS officer, one day to be accused of being responsible for the infamous "Malmédy Massacre" during the Battle of the Bulge, frowned. He wondered: Was von Rundstedt right? Somehow he felt the field marshal wasn't. He had a dark suspicion at that moment that "the Tommies would be back. They'd never give up."

Then the staff moved away to celebrate later with champagne and, for those young enough, elegant French whores imported specially from Paris for the great occasion. But the celebrations didn't change Peiper's dark forebodings. Not this month, perhaps not this year even, but one month in some year in the future, they'd come back. "We'd see them again. I was one hundred percent sure of that." And the future Obersturmbannführer Peiper, in two years' time the youngest commander of a regiment in the whole of the elite SS, was right that August evening. They would.

Back in England, the harassed medics, staff and planners were counting the cost in men and materiel. It was horrific. The Canadian Army had lost more prisoners to the Germans than they would in the whole eleven months of the campaign in France two years later or than in the twenty months of the fighting in Italy from 1943 to 1945.

While the propaganda war between Europe and Britain and America over the Dieppe fiasco waged back and forth, the Germans maintaining their troops had beaten off a full-scale invasion and the British stating that Dieppe had merely been a "reconnaissance in strength," a final tally of the casualties was made. It took several days to do so. Canadian losses alone totalled 215 officers and 3,164 other ranks out of just under 5,000 men involved. The 4th Canadian Infantry Brigade was virtually wiped out. In some regiments, casualties mounted to 67 percent. Of the seven major units involved, only the Fusiliers Mont-Royal, a French-Canadian outfit, managed to bring back its commanding officer.

One didn't need a crystal ball to know that the Dieppe raid had been a great failure. The more cynical among the survivors maintained that the 2nd Canadian Division had been sacrificed by Churchill to appease Stalin and Roosevelt. As one of them, Forbes West of the Royal Regiment of Canada, said bitterly afterwards: "We were sent into the raid largely to prove to our people, the Americans and the Russians, that the Second Front wasn't on. And we proved it!"

Higher up, Lord Lovat of the Commandos maintained the raid was "sign-posted all the way. We were expected to arrive at a certain date and we did. Where it mattered on the beaches, the enemy were waiting . . . The Canadians just weren't trained for the operation, whichever way it was done."

As for the Germans, they thought the Canadian planners—who had maintained before the raid that it was "going to be a piece of cake"—had been totally unrealistic in their expectations. The staff of the German 81st Corps, which defended the area (and would still be defending it two years later when the real invasion came), felt the British Commandos "were well-trained and fought with real spirit." Not so the Canadians: "On the whole [they] fought badly and surrendered in swarms." No doubt Canadian morale did sink to almost zero once the fighting men began to realize the truth of this "piece of cake" operation. As one prisoner told Hitler's interpreter Paul Schmidt: "The men who ordered this raid and those who organized it are criminals and deserve to be shot for mass murder!"

So what was the truth of the matter? Was the Dieppe raid ordered right at the top by Churchill to show his allies that France could not possibly be invaded in strength in 1942? Did the Dieppe attack have any value whatsoever? Were the lessons learned from it, if any, of use to the planners of the real invasion two years later?

Did the *New York Times* hit the nail on the head when it wrote one year later, on the first anniversary of Dieppe: "Someday there will be two spots on the French coast sacred to the British and their Allies. One will be Dunkirk where Britain was saved because a beaten army would not surrender. The other will be Dieppe where brave men died without hope for the sake of proving that there is a wrong way to invade"?

Mountbatten, like Montgomery, Churchill and the top Canadian brass, who were all equally compromised by the disaster at Dieppe, supported the view that the raid, although a failure, had shown the way ahead. He wrote long afterwards: "The Duke of Wellington said the battle of Waterloo was won on the playing fields of Eton. I say that the battle of Normandy was won on the beaches of Dieppe."

It was the line taken by most of the apologists for Dieppe. In truth, certain vital lessons were learned from that dreadful failure of 1942. The Normandy landings were accompanied by heavy naval bombardments unlike Dieppe. Allied planning was much more intricate and close. Above all, no attempt was made to capture a fortified, German-held port. Instead, the Allies hit broad, open beaches and took their own port with them, the British-conceived Mulberry harbor . . .

Still, were those innovations worth the terrible casualties? I think not. Planners who work in the comfort and quiet of their offices have not always needed to spill the blood of young soldiers simply to test their theories. So why did the Canadians and the British land at Dieppe that August so long ago; star-crossed, doomed men marching unwittingly to their inevitable fate?

There can be only one answer to that overwhelming question. The Canadians *had* to fail! We can guess that it was Churchill who gave the direct command that the raid be carried out, whatever the consequences. But who put the pressure on Churchill and subsequently the Canadian government to carry out a command that even a military layman could see would end in failure?

The answer is obvious. There was only one man more powerful than Churchill in the Western World. He had just brought a reluctant United States into the war against the Axis powers. Now he was attempting to force his will upon the American people—and on his own top brass. Many of his admirals and generals wanted America's first priority to be given to the war against Japan. He thought differently. Europe was more important, in his opinion. But it was obvious that a major invasion of the Continent could not be staged in the year 1942. The American people had to be shown what would happen to a large-scale attack against France. The result: Dieppe.

We shall never know if it was Roosevelt for certain. There will always be a certain mystery about the events at Dieppe that tragic summer. But, as a great American writer, doomed himself, wrote long before, "So we beat on, boats against the current, borne back ceaselessly into the past . . ."

13

First Chindit

BY GEORGE F. CHOLEWCZYNSKI

By mid-1942, the success of the Japanese armed forces seemed to auger the zenith of a vast militaristic empire that encompassed all of Southeast Asia. In the months after Pearl Harbor, the Japanese had thrown the Americans out of the Philippines, the Dutch out of the East Indies, and the British out of Hong Kong and Malaya. When Singapore fell, over 85,000 British and Commonwealth troops went into the nightmare of Japanese captivity, in the largest and most humiliating surrender in British history. The Japanese then drove into Burma, and by June had managed to push the last Commonwealth soldier beyond the Chindwin River. Thousands of soldiers and civilians had died trying to reach the safety of India.

The Allies had not only lost Burma's vast natural resources, but their only viable supply route to China had been cut. British morale was at a low ebb. The Japanese soldier, previously an object of ridicule, had proved himself not only to be skillful, but also ruthless beyond imagination. No one had expected that Europeans would ever face the treatment meted out to the Chinese that they had seen in newsreels years before. It was a shock when the fate of prisoners, the wounded and women in the hands of the Japanese was revealed. British soldiers in Asia realized that they could not expect any form of mercy from their enemy and, among many troops, the belief spread that the men of the Rising Sun were well-nigh invincible.

The British Army had attempted offensives in Burma in the coastal Arakan region in September 1942, and again in early 1943. These had been dismal failures, and resulted in not only heavy British casualties but also the reinforcement of the Japanese Army in Burma by several divisions. Not only had their efforts led to nothing but more losses, but now there was real fear that the Japanese would eventually threaten India, the jewel in Britain's imperial crown.

The last British offensive had followed a master plan for the reconquest of Burma named "Anakim." Part of Anakim called for the Burma Rifles to infiltrate into the hills of northern Burma and set up a foundation for armed resistance to the Japanese by the mountain tribes who lived there. Due to the failure of the recent military operations, however, Anakim was canceled. The Burma Rifles, known colloquially within the British Army as the "Buriffs," was a regiment of native soldiers led by the British colonials who ran the plantations, railroads and commerce of the colony—and they were one of the very few military formations that had not been shattered during the Japanese invasion. They were soon to be inserted into Burma as part of a diversionary raid led by a brigadier named Orde C. Wingate, an individual who was viewed by the troops under him as a Messiah, and the officers above him as a madman.

Wingate had many curious habits, and most would not endear him to anyone. His personal hygiene was beyond description, as was the state of his uniform. He would be seen frequently munching on an onion, the juices of which would merge with all the others that had soaked into his bush coat from other meals he had eaten over the past weeks. When he spoke, his words were carefully chosen and delivered with a rasp that was a result of a suicide attempt in Cairo in 1941. Wingate would speak with his eyes either locked on the horizon or burning through his audience. His speech would be a melange of scholarly discourse, classic philosophy, biblical quotations and treatises about long-range penetration and guerrilla warfare. Those who listened to him and did not dismiss him as a lunatic found themselves hypnotized by the strange bearded man in the smelly uniform.

Wingate had been born in the Himalaya Mountains in 1903, the son of a British officer and the grandson of the former British High Commissioner for both Sudan and Egypt. His family were members of a religious sect known as the Plymouth Brethren. Their beliefs had barely changed since the time the majority of the denomination had emigrated to North America in 1620. Brought up in an atmosphere of psalms, proverbs, chapters and verse, Orde Wingate had a difficult time in public school, and during his subsequent studies at the Royal Military Academy, Woolwich. Commissioned an artillery officer, he followed the path of a distant cousin, T.E. Lawrence, and studied Arabic language, customs and culture. Posted to Sudan in 1928, his reputation as a loner and oddball grew among his fellow officers as Wingate would spend long periods in the desert, denying himself food and water to test his powers of endurance.

In 1938 Captain Wingate was transferred to Palestine, where he took up duties as an intelligence officer. The biblical scholar was pleased with his assignment, though it was considered a career dead-end in the British Army. He and his wife furnished their home so that it appeared no different from any other native dwelling in Palestine, and they ate the local food. Visitors were treated to unusual evenings that were divided between listening to Christian religious music on the phonograph and discourses on Jewish religious literature by Wingate. All this further alienated him from his fellow officers.

Though his initial sympathies were with his beloved Arabs, Wingate soon noted the plight of the Zionists, who were trying to return to their Holy Land. Any support or sympathy that the British government or army had had for them had vanished long before.

The Zionist farm colonies were under attack by Arabs. Their women and children spent the nights in bunkers while their men patrolled their fields and orchards against marauders who sought to kill livestock and destroy the crops. Wingate was outraged at the plight of these hardworking people trying to construct a life for themselves in the desert. His chance to help came when the marauders also disrupted the British petroleum pipelines.

Wingate offered a solution by the creation of a small force, "Special Night Squads," to bring peace to the area. These units would be composed of two hundred British soldiers and four hundred local Jews. The British commander in Palestine, General Archibald Wavell, though that this solution to level the playing field between the Arabs and Jews had merit, and would eventually cause fewer problems for the British.

Initially, the Zionists were confused by the strange man who said that he would like to help them to help themselves. That he was a British officer was bad enough, but the fact that he was an intelligence officer aroused suspicions. But his scholarly knowledge impressed the local religious leaders. As word of the rabbis' respect for him spread through the community, Wingate soon had volunteers for the Special

Night Squads. These began a cycle of patrols that involved brutal fighting where no quarter was given or expected by either side. Wingate himself was wounded on two occasions. Soon Zionists from all over Palestine were serving with the organization. When they returned to their homes, they shared the knowledge of guerrilla warfare that they had learned. This was the unlikely beginning of the Hagganah.

When not serving either the British Army or the Special Night Squads, Wingate reveled in the military history of the Old Testament. One of the few British officers who would meet with Wingate socially was a peppery cavalry officer who studied the classics and attended local universities wherever he was posted. Together they would go to the battlefields where Jonathan, Gideon, Saul and David had fought, and they would analyze scripture and discover the old battle lines. Yet there was another purpose to these intellectual exercises besides social and mental stimulation: the cavalryman was assigned to the Trans-Jordan Frontier Force, and the two scholars ensured that their relative military units would avoid conflict through a congenial exchange of intelligence. By the time the cavalry officer finally retired from the British Army to take a position as rector of London University, General Sir John Hackett had commanded a parachute brigade at Arnhem and had been NATO's northern commander-in-chief.

Wingate was decorated with the DSO for his efforts, and was recalled to Britain. His relations with his fellow officers had not improved, and grew worse when Britain issued a White Paper forbidding further Jewish immigration into Palestine. The outbreak of World War II found him assigned to London's anti-aircraft defenses. During the Blitz Wingate was unhappy with the abysmal performance of the anti-aircraft defenses and filed written protests, then resigned his commission. When Italy joined the war and threatened Egypt, Wingate, with the blessing of Chaim Weitzmann, began lobbying the British government to raise a Jewish Legion to fight the Nazis. Wingate wanted to use his expertise and knowledge of the desert to attack Italian supply columns behind the lines. His proposal was considered, and then refused.

At this point Wingate's former commander, General Wavell, had a proposition for him. Wavell was one of the few high-ranking officers who could not only stand to be in the same room with Wingate, but also listened to what he had to say. Wavell was thinking that Wingate would be the perfect person for a backwater campaign in Abyssinia.

Italy had invaded Ethiopia in 1936. Opposed by spear-carrying tribesmen, Mussolini's troops overwhelmed the country with tanks, modern artillery, bombers and poison gas. The deposed emperor, Haile Selassie had aroused much sympathy in the world, but little else. Sudan was inundated with refugees from Ethiopia, and Major Wingate returned here to create an army and restore the "Lion of Judah" (as Selassie was known) to his throne. There were many volunteers, but Wingate was selective and only one out of twenty was accepted into the ranks of his "Gideon Force."

Strengthened by a battalion of Sudanese, Wingate led his poorly armed, but eager and well-disciplined Ethiopians back into their country. The local population provided excellent intelligence, and the Italian defenders little resistance. Where there was resistance, the Ethiopians showed no mercy. Prisoners were released after their weapons were turned over to equip the invaders, and they spread panic to the neighboring garrisons. Outnumbered three to one, the Gideons managed to bluff the Italian garrisons into surrender, one after another. When Haile Selassie triumphantly re-entered Addis Ababa, he was escorted by Orde Wingate mounted on a white horse.

The triumph was short-lived for Wingate. He was recalled to Cairo, where he again encountered scorn from the officer corps. He was suffering from malaria, and the doctors who were treating him began to question his sanity. Ordered to shave, Wingate returned to his cheap hotel, took his bush knife and drew it across his throat. He was saved from death by a person in an adjoining room, and from oblivion by General Wavell.

General Archibald Wavell had been transferred from the Middle East to India to assume the position of Commander-in-Chief, Far East. As part of his Arakan offensive, General Wavell envisioned a long-range penetration operation to precipitate chaos and confusion along the Japanese lines of communication in central Burma. Having seen Wingate in action, Wavell was confident that he knew the right man for the job.

Orde Wingate arrived at Wavell's headquarters in India as the last of the retreating British soldiers staggered across the Chindwin. The first thing he did was request any and all books about the tactics and training of the Japanese soldier. This was later expanded until Wingate had consumed every available work on the customs of both the Japanese and the Burmese—books about the climate and geography of Burma, as well as every after-action and intelligence report available from the recent disastrous campaign.

In a matter of weeks he laid out his proposals to Wavell. Wingate told his commander that although the Japanese soldier was an excellent jungle-fighter, he was not born that way, and the Japanese homeland no more in tropical jungles than the English midlands. Wingate asked for 3,000 fit men, and stated that after a few months he would make every one of them capable of coping with the best soldiers the Japanese Empire could put in the field.

In 1942, however, the British Empire was hard-pressed on all fronts. The Germans were winning the Battle of the Atlantic, bringing the United Kingdom face to face with the prospect of starvation. The disaster at Singapore had been echoed by another huge capitulation of Empire troops in North Africa, at Tobruk. The Allies were faced with crises on every side, and the British Army in India had the absolute last priority for manpower, or for any kind of supply or equipment.

The military formations in India were either licking their wounds, protecting the frontier, or engaged in internal security duties. The one unit that could be spared was the 77th Indian Infantry Brigade. It consisted of the 13th King's (Liverpool) Regiment, the newly raised 3/2nd Gurkha Rifles and the veteran 2nd Burma Rifles. The 13th Battalion of the King's Regiment had been raised after the fall of France, had arrived in India in January, and had served on internal security duties. The men were conscripts from the grimy cities of Manchester and Liverpool, many of them married or middle-aged, and often cynical by-products of the industrial revolution. Wingate did a thorough combing-through of the unit, and almost half were transferred out for poor physical condition, and replaced by fit men from other units. A commando company, the 142nd, was raised, accepting volunteers from all units and services. Wingate asked for and received RAF personnel, mostly signalers. The entire ad hoc command then moved to the jungles near Patharia in central India to begin their training.

Wingate began by reorganizing his force. The new structure of this unorthodox formation would have a brigade headquarters, and the remainder would be divided into two groups. Instead of conventional infantry companies, the basic unit was known as a column. Each column would consist of three rifle platoons, a support platoon with mortars and machine guns, a mixed platoon of commandos and sappers from 142nd Company, a platoon of Burma Rifles to act as scouts and interpreters, and an RAF signal section. No. 1 Group would consist of Columns No. 1 through No. 4, and was primarily composed of Gurkhas, while the King's Regiment would provide the majority of the personnel for Columns No. 5 through No. 8. All officers were to develop basic vocabularies of both Urdu and Gurkhali.

During training it was emphasized that the jungle was neutral, but could be turned into an ally, as the Japanese had proven. The Brigade would live under canvas in the jungle rather than in conventional barracks. The first few nights were difficult for the uninitiated. Sleeping under stuffy mosquito nets, the noises of jungle insects and animals proved to be louder than the night sounds of a seaport or mill town. Jackals, panthers and cobras would be frequent visitors.

The personal equipment that the British soldier carried was reviewed. British web gear was designed for interchangeability and ease of manufacture, and ergonomic factors were given little thought. New "Everest" rucksacks replaced the cumbersome traditional large and small packs. The steel helmet was discarded as having little protective value compared to its weight, and replaced by the Australian style bush hat. The khaki Indian-manufactured cellular cotton uniforms were dyed green. In addition to the normal load of ammunition, each man now carried either a Kukri or machete as part of his kit. All were to carry a groundsheet, mosquito net and a pair of canvas, rubber-soled boots for when their sturdy leather boots needed a chance to dry.

Brigadier Wingate devised a ration system that would replace traditional bully beef. Besides being heavy, the cans of fatty meat were unpalatable in the jungle heat and humidity. The daily ration for each man would now be twelve hardtack biscuits, 2 ounces of raisins and nuts, 4 ounces of dates, a bar of chocolate or hard candy, vitamin C tablets, and a packet of powdered milk so that they might have a proper cup of tea. Each soldier was expected to carry six days' rations in his rucksack. It was anticipated that air drops would be expected on the average of every five days to replenish rations, ammunition and medical supplies.

To transport heavy weapons, the RAF radio sets and other unwieldy equipment, mules would be used. Several bullocks would accompany each column, and though poor load bearers, they could be of limited use in this capacity until slaughtered to supplement the soldiers' rations. Officers would have a cavalry horse.

Mules, one thousand of them, arrived in November 1942. Many had made the long voyage from the United States via South Africa, and arrived in poor physical condition. Not only that, but they had not been trained or even broken. It was left to men who had been born on the Mersey or in the Himalayas to do this. It took several weeks for the mules to regain their strength, and for Liverpudlian, Gurkha and Missouri jackasses to come to a mutual understanding. After initial difficulties, the Brigade officers were amazed how both the British city dwellers and diminutive Gurkhas learned to train, pack and lead their stubborn charges.

Wingate would make frequent appearances. He would gather the men and speak to them, and the effect was hypnotic. The sarcastic conscripts were spellbound by the bearded man in the sun helmet. Despite the hardships and exhausting training, they felt they were an integral part of what was going on. Military nonsense was at a minimum. The barriers between the officers and the ranks became thin as all lived, worked and endured the hardships together.

After several months of organization and training, the Brigade began field exercises. No longer did the men have the luxury of sleeping under canvas. Wingate drove his officers and men mercilessly. The columns maintained contact with each other, but operated separately, as they were expected to do once they were behind enemy lines. They now lived in the jungle on an operational level, marching for forty miles, and then assigned to blow up a guarded "railroad bridge" and escape rather than bivouac for the night. The skill most heavily emphasized was jungle navigation.

This period of final training coincided, unfortunately, with the arrival of the monsoon season, and the soaking wet soldiers had to endure not only the jungle but also the wretched weather. It was during these field exercises that it was learned that the mules not only loved to eat the leaves off bamboo plants, they thrived on them. This happy discovery meant that large amounts of fodder that had been calculated into the weight and scheduling of air drops could now be replaced with ammunition, food and, best of all, mail from home.

Despite the fact that Wingate was unable to obtain the best men, he had whipped his long-range penetration force into an admirable unit. The Brigade adopted the name "Chindits," after the statues of dragons that guard the entrances of Burmese temples. Morale was high among the British components. As for the 3/2nd Gurkhas, despite the fact that they were a green unit, all of these men had learned what was expected of them at the hearth, from their parents, grandparents and uncles, and it was inconceivable that they would fail.

At the end of January 1943, Wingate declared his men ready, and met with General Wavell to discuss his future mission. Because of the failure of the Arakan offensive, Wavell considered the Chindits' mission into Burma pointless at this time. Wingate would not be pacified, and he convinced Wavell to let him go, regardless. Wingate outlined a plan by which his force would set out to disrupt the enemy's communications by cutting the strategic road and rail links that ran north from Mandalay to Myitkyina. He hoped that this would cause great confusion in the enemy's rear, and looked forward to leading the Japanese on a wild goose chase. Wavell acceded.

Early in February, the Chindits were assembled at Imphal. Wingate addressed his men, and told them that should anybody be wounded or stricken with disease, they would have to be left behind because there would be no way to evacuate them. He did say that he would try to leave them in the care of friendly Burmese villagers, but that this would not always be possible. Every man knew what mercy they might expect from the Japanese. At this late date, Wingate announced that any man who wished to leave the Chindits could now do so. Few men chose this option, because most had become confirmed followers of their strange messiah.

The Chindits neared the Chindwin River after a five-day march from Imphal. To the onlooker it would have seemed an odd parade of officers on horseback leading men in floppy bush hats—mules, bullocks and dogs happily trotting alongside. Each column had a dog that had two handlers; the handlers would be stationed at opposite ends of the column, and the dogs would carry messages between them. In front of them all was the bearded man in a sun helmet and stained bush jacket who had chosen a pony as his personal method of transport. Supply drops would be made by the RAF, using the Dakotas of the 31st Squadron and by Hudsons, stripped of their gun turrets in order to carry more cargo, of the 194th Squadron.

The Brigade divided into two groups. No. 1 and 2 Columns would cross the Chindwin south of the main force, and provide a diversion. A smaller party led by Major John Jeffries, attired in the uniform of a brigadier, down to the scarlet collar tabs, would cross even farther south. Jeffries' purpose was to reconnoiter and gather intelligence about the Japanese; he would cause a further diversion by spending money fast and freely, buying large quantities of rice for an army that would never arrive.

The northern group melted into the jungle during the day, and marched by night. On the morning of February 15, 1943, the Chindits reached the Chindwin. In front of them, the 400-yard-wide river reflected a light emerald color in contrast to the savage green jungle that extended to the horizon beyond.

The brigade spent the day in the jungle. Patrols were sent up and down the shore trying to obtain boats. The crossing began at dusk.

The Chindits' operational area (box), 1943.

22. Gunther Prien's daring raid on the British fleet base at Scapa Flow created renewed enthusiasm for the U-boat arm among Germany's high command.

23. A type VII boat of the kind Prien commanded. Note the narrow beam; there was little comfort or safety built into these stealthy craft.

24. The port of St. Nazaire, with the Old Mole at far right center, and the *Forme Ecluse* beyond it angling into the inner basin.

25. HMS *Campbeltown* after being rammed against the dry dock. Note the gashes torn in the bow by coastal defense guns and the German soldiers on deck trying to puzzle out the intention of the action. They didn't have long to wait for the answer.

26. Canadian troops training on the Isle of Wight for the raid on Dieppe. After their voyage to England they had endured years of training and inactivity before being tapped for a major operation.
27. Below, wooden landing craft head for the French shore.

28. Burning landing craft, disabled Churchill tanks and Canadian casualties give mute testimony to the difficulty of taking a heavily fortified beach near a major port.
29. Below, some of the survivors of Dieppe who made it back to England. The operation has remained controversial to this day.

30. Chindits pause to make camp somewhere behind the Japanese lines in Burma. They have discarded steel helmets in favor of Australian-style bush hats to fend off the sun, the rain and the leeches that drop out of the trees.

31. Chindits in Burma, with a recalcitrant burro and a bottle of wine.
32, 33. Below, two aspects of Orde Wingate (1903–1944), the biblical
scholar who was also a dynamic, innovative leader of troops. The photo
at right was taken several weeks before his death in a plane crash.

34. B-17 Flying Fortresses fly steadfastly through enemy fire to their target.
35. Below, German pilots pose atop a Messerschmitt 110 Destroyer; note the rocket tubes placed beneath the wing.

36. By 1945, Otto Skorzeny (1908–1975) was well known to the Allies. Thousands of these uncomplimentary posters were distributed behind the American lines.

37. He wasn't considered "the most dangerous man in Europe" for nothing. Here Skorzeny squeezes himself into a small plane behind Benito Mussolini, despite the advice of his pilot.

38. Otto Skorzeny expertly conceals his tension as he awaits his trial at Nuremberg.

39. Captain Donald R. Strobaugh (left) with young Belgian ParaCommandos on Ascension Island, three days before the Stanleyville jump.

40. A ParaCommando checks his gear at Kamina airfield.

41. Although the United States declined to use its own airborne troops in the Congo, the American presence was still detectable to sharp-eyed observers.

42. Disarmed Simba prisoners being questioned by ParaCommandos shortly after seizure of the Stanleyville airfield.

43. Right, rescued hostages being evacuated from Stanleyville.

44. Below, Major Roger Hardenne of Para Regiment HQ and paratroops loaded up on motor-tricycles.

45. An Israeli naval commando fires a grenade from a modified AK-47 rifle.

46. Members of the secretive Flotilla 13, the IDF's naval commando unit, prepare for SCUBA training, skills that were put to the test in their assault on Green Island.

47. Colonel Arthur D. "Bull" Simons briefs the Son Tay raiders just before their departure by helicopter from an air base in Thailand. Simons began his raiding career in WWII and capped it by spiriting a handful of hostages out of Iran in 1979 in a private rescue operation.

48. Captain Richard J. Meadows, leader of assault team "Blueboy," whose helicopter entered the Son Tay POW camp in a controlled crash that took the main rotor off on a tree. The megaphone in his left hand is for informing prisoners that American soldiers have arrived and that they should stay down.

49. A group of Son Tay raiders on board one of the helicopters inbound for the POW compound in North Vietnam.

50. After the raid into Tehran, four-engined C-141Bs were intended to land in the desert and transport the American hostages to safety.

51. U.S. Navy RH-53D helicopters aboard the carrier *Nimitz* preparing to embark on the raid into Iran. Flown by Marine Corps pilots, they were to be plagued by dust storms, hydraulic system failure and a lack of night vision equipment that would hamper the mission.

52. While Delta Force rescued the American hostages, AC-130 Spectre gunships were to orbit the embassy compound, laying down enough fire from their 20mm, 40mm and 105mm cannon to convince angry crowds or Iranian reinforcements to keep their distance.

53. Part of the wreckage at Desert One after the raid was called off. In the years since 1980 the United States has learned from its mistakes and, with the implementation of Special Operations Command, no longer has to reinvent the wheel in order to launch swift and coordinated raids.

There was a full moon that night, and it shined so bright that the veterans of the expedition swore they could read a newspaper by it. Though they had been trained in water crossings, the mules had never seen such a large body of water. Some of them managed to make it almost to mid-stream before they turned back for the safety of the west bank, their cursing handlers in tow.

The result was a good deal of confusion on the banks of the Chindwin as the subsequent columns emerged from the jungle. The track to the river was steep and muddy. The situation was not improved when the moon slipped below the horizon, as the crossing continued in the black jungle night. When dawn broke, half the Chindits were still on the west bank of the river. New crossing sites were scouted out, and on the following night the transit was completed.

Two days before the Chindits reached the Chindwin, a party on horseback led by RAF officers had crossed the river, and proceeded to a point ten miles beyond. As they traveled, they canvassed the Burmese villagers as to the presence of any Japanese. They were told that no patrols had been in the area for at least a week. The party went on to their destination, and arranged an air drop. The columns of the northern force soon rendezvoused there and collected supplies. Wingate was happy that this, the first supply drop of the raid, was successful.

The southernmost element of the Chindits, under Brigadier Jeffries was operating out in the open. He enjoyed his role, walking into known pro-Japanese villages and haranguing the Burmese headman in his best "brigadier" manner about the imminent return of the British Army. After a few days of deception, Jeffries' men ambushed a Japanese patrol. Jeffries estimated that it was twenty strong. With no loss to themselves, the Chindits killed seven of the enemy. Then Jeffries immediately changed direction and headed south, rather than east, to draw off the Japanese who would surely come looking for them.

The Japanese began massing troops near the lower Chindwin and started to hunt for the elusive British. Jeffries' deception party traveled without any supplies, and on occasion were fed by the Burmese villagers. Their biggest problem was water. After leading their pursuers on a wild goose chase for a week, Jeffries and his party melted into the jungle and, moving in a northeasterly direction, went off to rejoin the other Chindits.

The northern group continued to push eastward, towards the Mandalay-Myitkyina railroad. The Chindits found that their maps were not always accurate. Instead of areas showing solid ground, the raiders and their mules found themselves plodding through swamps. The men found themselves hacking down small trees to build make-shift causeways. They were now in the middle of the jungle, and constantly on the alert for the enemy.

Progress was slow. Each of the three rifle platoons in each column would rotate as perimeter guard. That meant they would have to stay alert the entire night while the others managed to get a few hours' sleep before pushing off the next morning. Traveling through the jungle exposed the men to intense labor as they had to hack trails in the heat and humidity—tormented by lice, mosquitoes and ticks. At night the temperature would plunge, and the Chindits would shiver, wrapped in the single wool blanket that was provided for each man. The next morning it was wake up, load up the mules and move out. Brigadier Wingate forbade the brewing of tea in the morning as "a waste of time."

Now the Chindits were finding signs of Japanese patrols: footprints left by Japanese boots were being discovered with greater frequency. Major Calvert's No. 3 Column discovered a recently abandoned enemy campsite, along with an elephant and its Burmese handler, which had been left behind. Calvert welcomed the reinforcement to his pack column while the Chindits brewed tea in the still warm ashes of the campfires.

The columns bivouacked near each other on the night of February 24, when a second, and massive, supply drop took place. It took three days to receive and distribute the supplies. The Chindits' morale was good, and was supplemented by twenty bottles of rum.

On March 2, 1943, the Gurkhas of No. 2 Column were on their way to blow up the railroad near Kyaiktan, 40 miles south of Nankan. They had made it to their destination and were placing the charges when a train stopped outside the village and started disgorging troops. The Gurkhas were almost immediately under mortar fire as Japanese infantry swept wide around both sides of them. Then Japanese machine guns opened up. The mules, terrorized by the explosions and tracer bullets, stampeded, bowling over Gurkhas as they tried to set up defensive positions. The Chindits maintained their composure, and broke up into small parties in order to take on the Japanese. The antagonists became intermixed in the dark jungle night. The fighting was vicious and hand to hand.

By dawn, the Japanese thrust had been blunted. Then the column, under sniper fire throughout the day, tried to make order out of the shambles. Gurkhas attempted to salvage what they could from the dead mules and what these had scattered through the surrounding jungle. In addition to many mules killed and missing, the column lost many of its heavy weapons in addition to precious radio sets. After dark, No. 2 Column managed to set off its demolition charges on the railroad line before slipping back into the jungle.

A few days later, No. 2 Column managed to link up with No. 1, which had managed to cut the railroad in two places without confronting the enemy. Though No. 2 Column had suffered casualties, and heavy losses in animals and equipment, the diversion by the southern force allowed the majority of the Chindit expedition to continue relatively unmolested.

As the northern group pushed eastward, No. 4 Column was ordered to divert further enemy strength with an attack on the village of Pinbon. The Chindits met the Japanese on March 3, 1943. The fighting was sharp, and followed by a series of vicious skirmishes. After the initial reports of contact with the enemy, Wingate heard nothing further from No. 4 Column. It was as if it had been completely swallowed by the Burmese jungle.

The rest of Wingate's force pushed on towards the Mu Valley and the railroad. No. 3 Column was assigned to attack Nankan Station, clear it of Japanese and destroy the railroad facilities. The column was led by a tough, lantern-jawed major of the Royal Engineers, J.M. Calvert. He had already spent much of the war behind enemy lines, and had been one of the last British soldiers to leave Norway. In both Malaya and Burma, he had stayed behind the retreating armies to perform rearguard demolitions, destroying anything of value to the enemy. His nickname was "Dynamite Mike."

When Calvert and his men reached the station, they found no Japanese, and quickly began their demolition preparations. As the last charges were being set, two trucks loaded with Japanese soldiers appeared. They were ambushed by Burriffs, but several more truck-loads of enemy infantry arrived on their heels. The Burriffs managed to hold their ground until the demolition charges went off.

Over 75 charges had been placed. No. 3 Column had destroyed six miles of railroad and two railroad bridges, including their abut-ments. Now, Burriffs slowly had to give ground: Japanese attackers were in the village of Nankan. Under machine-gun fire, two of Calvert's men began laying charges on yet a third railroad bridge. They lit the fuses and, pursued by bullets, dashed for cover as the bridge blew up.

It was at this point that one of the demo men started shouting to Calvert. He pointed out a group of some fifty Chindits coming out of the jungle. They had been separated from their column for three days, lost and living on palm hearts and roots. Major Calvert was as relieved to see them as they were him. With the reinforcements Calvert set up his mortars and started shelling the Japanese in the village. The enemy retreated into the jungle, leaving behind a dozen dead. Calvert then had his men booby-trap the road.

When Calvert assembled his men before they pushed off, he dis-covered that his column had suffered no casualties. Calvert was a happy man. He had accomplished much this day, which was his thir-tieth birthday. He received his present seven months later when the DSO awarded for his actions at Nankan was gazetted.

To the north, Major Bernard Fergusson's No. 5 Column was detailed to blow up the railroad bridge over Bonchuang gorge. The column had marched 300 miles since it crossed the Chindwin, break-ing only to gather air drops. Major Fergusson was a Scot, commis-sioned in the Black Watch. He was polished and well educated, and a frequent contributor to both *Punch* and *Blackwood's* magazines. A veteran of the fighting in Libya, he had given up a lieutenant colonel-cy on the General Staff in Delhi to lead a column into Burma. On the morning of March 6, Fergusson moved out from his bivouac on the last leg of his trek to Bonchuang, but became involved in a skirmish with Japanese in a village enroute. Sixteen Japanese bodies were found in the aftermath, but No. 5 Column had suffered two killed and seven wounded. Five of the wounded, including one officer and three non-commissioned officers, were hurt too severely to be moved. The dread-ed question of what to do with the wounded finally arose. Reluctantly,

Fergusson left them in the shade of one of the houses with their packs and extra water.

The column reached the bridge and had it ready to blow soon after dark. As the column moved out of the area, a flash illuminated the men and mules as the entire countryside shook with the blast. Major Calvert and his men heard the explosion and saw the flash ten miles away.

The Chindits so far had had a very successful mission. They had cut the railroad in thirty places, and caused pandemonium on the Japanese lines of communication between northern and southern Burma. The Japanese were confused about what exactly was going on behind their lines. Scouting parties were sent out to discover the size and nature of their enemy. Many did not return, which heightened the bewilderment. Those that managed to engage the British before they could break contact and disappear into the jungle gave reports that greatly exaggerated the strength of the Chindits operating behind their lines.

Wingate followed the tactics he had developed in Palestine and Ethiopia. Relying on deception, he would avoid combat whenever possible, and attack only when and where the odds were strictly in his favor. After the element of surprise had run its course, he would withdraw. Whenever possible, the Chindits would evade, rather than confront, the enemy.

After attacking the railroad, the Chindit force was in relatively good shape. One column had disappeared and another had been mauled but, after nearly a month in the jungle, the force was still strong and morale high. However, Japanese efforts to find the raiders had been stepped up. The villages in the area were poor, and could not be depended on as a source for food. In one, No. 5 Column was spotted by a Japanese airplane. It circled back, but instead of bombs, dropped leaflets. The assurances of good treatment if the Chindits surrendered provided a rare opportunity for laughter among the raiders, and were then put to good use as toilet paper.

At one point, there was no place or opportunity for the Chindits to receive a badly needed air drop. Brigadier Wingate sent orders to all columns that the next drop would be near the village of Pegon, twenty miles east of the Irrawaddy on March 12. The supply situation was so critical that No. 3 Column's radio battery went stone dead after confirming receipt of the message.

All units started on their way east, changing direction temporarily whenever spotted by the Japanese. There were a few sharp skirmishes before the Chindit columns reached the Irrawaddy. It was the dry season, and the wide river had broad sand banks on both of its shores. While Major Calvert was contemplating how to cross the river unobserved, a miracle happened. Three Hudsons appeared overhead, leaving a stream of parachutes in their wake. The tough major, in his own words, stood there and wept like "a silly old lady."

The Chindits got across the Irrawaddy with the help of boats loaned by the Burmese. The east bank proved to have thick growths of elephant grass. Calvert moved south, and met No. 1 and 2 columns. On March 16, Calvert and Major Jeffries, who had by now abandoned his charade as a brigadier, celebrated St. Patrick's Day with a libation distilled from the raisins in their rations. It was a day to celebrate—all of Wingate's remaining men were now east of the Irrawaddy.

The land beyond the Irrawaddy was far different from what the Chindits had been operating in. Instead of jungle, the area was covered with teak plantations that were devoid of cover or underbrush. There was no concealment from the Japanese, and the blazing sun baked man and mule. In this dry season there was little water, and not even bamboo groves where a precious mouthful might be obtained from the stalks.

As the columns headed towards their rendezvous they were running into more and more Japanese. Encounters were frequent, and as the columns crossed each other's paths they found bivouacs with prints of both British and Japanese boots. As the columns eventually made contact with Wingate, he greeted them with the news that the Chindits were ordered to return to India. The promised supply drop was made, and the men again had rations, clean uniforms, gasoline to recharge their radio batteries and a new pair of boots for each man.

But the Japanese were closing in. Wingate issued his orders for the withdrawal. He emphasized that the men were more important than the equipment. The columns made their last arrangements for supply drops, and then destroyed their radios. The Chindit columns then divided out the remaining mules and broke into small parties, heading out in all the directions of the compass to confuse their pursuers before they turned west, back toward the Chindwin River.

The weeks went by as the Chindit columns broke into ever smaller groups to avoid the Japanese. Hunger, thirst and disease stalked them, as more and more men succumbed to fatigue after their long weeks in the jungle. Between the Chindwin and the Irrawaddy, freak rainstorms turned the small creeks into raging torrents. When the men could stop, the remaining mules were killed, usually by cutting their throats so that the Chindits would not be betrayed by a gunshot. Those groups who had the time cooked and ate these beasts which had served them so well. Native to the region, the Burma Rifles ate jungle roots and snakes; the men from Liverpool and Manchester soon followed their lead. Tortured by dysentery, malaria, jungle sores and insects, finally handfuls of emaciated survivors began reappearing on the bank of the Chindwin. It was now the third week of April, and there was still no word about what had happened to Brigadier Wingate.

Wingate had reached the Irrawaddy during the early hours of March 29. As he had predicted, the Japanese had collected most of the Burmese villagers' boats. A few dugouts were found, and Major Jeffries was ordered to supervise the crossing. As the first dugout pushed off, Jeffries saw small fountains of sand erupt in front of him. The exhausted officer then realized that he was in the sights of a Japanese machine gun. Taking cover behind a sand embankment, he saw a large contingent of Japanese moving along the opposite shore, some of them pausing to set up a mortar. Wingate ordered everybody back into the jungle, and the Chindits set up their mortars and returned fire.

Wingate himself walked calmly back into the jungle despite the bullets and mortar fragments flying around him. He sent orders to all the officers to assemble as soon as possible. He had everybody break into small groups, then move east until they had lost the Japanese. Then everyone was to attempt to make their way out of Burma on their own.

Eight days later Wingate again tried to cross the Irrawaddy. By now, every remaining boat on the banks of the river had been smashed by the Japanese. The Chindits in Wingate's party tried to repair them, but the enemy had been thorough. They retreated into the jungle, judging that they might have better luck farther south. At the new crossing site the following day, they spotted Japanese on the opposite bank of the river. Wingate's party again withdrew into the jungle for the night. The next day, they found a heavy dugout canoe on a lake and dragged it to the river. To their relief, there was no sign of Japanese on either shore.

In the middle of the river, to everyone's amazement, they saw a boat heading upstream. Wingate hailed it and discovered that the Burmese peasant had permission from the Japanese to carry the corpse of a villager upstream. The peasant had taken the opportunity to bring a cargo of salt and tomatoes with him. The food was quickly purchased, and consumed, and a sack of silver rupees convinced the boatman to dump the corpse and provide ferry service for the Chindits.

Three-quarters of Wingate's group had crossed the river when they were discovered by the Japanese, who brought the enterprise to an end with rifle fire. Wingate watched as the nine Chindits remaining on the east bank disappeared into the jungle.

The messiah drove his remaining group of followers mercilessly. He would not even allow them to hold up the group by letting them stop when they were ravaged by diarrhea. He told them bluntly that they should not worry about their trousers, that their stopping not only slowed the group, it also weakened their own will to go on.

As the group approached the Chindwin, Wingate sent out his Burriffs to scout the area. They returned to report that every village they had encountered had Japanese in it, and all boats had been commandeered by the enemy. One man had managed to buy some rice in one of the villages, and obtained information that there were Japanese patrols throughout the area looking for them; the Japanese had just combed the area where Wingate's force had been the day before. Wingate decided that this would be the best place to stay until they tried again for the river. He again sent out the Burriffs to find boats, but they never returned. They were captured and executed by the enemy.

The group evaded the Japanese by marching and counter-marching through the area. Wingate studied his maps, and decided to make for an area where the west bank of the Chindwin was regularly patrolled by the British. He picked four men to go with him. All were strong swimmers, and they would try to cross the river and contact a patrol before returning for the rest. After an agonizing march through elephant grass, the party reached the river on the afternoon of April 28. The weakened men waded into the water. They wanted to make the defiant gesture of returning with their packs and rifles.

They struggled across, and despite the fact that not everybody's pack, rifle and boots made the crossing, all five men did. They met a peasant who led them five miles to a British Army post. While enjoying their first tea in weeks, they told their British officer hosts about the remaining men on the other side. A patrol with enough boats to pick up the others was mounted, and went off to the river to wait for a signal. They waited the entire night, and there was none.

After Wingate's departure, Japanese patrols had forced the rest of his party away from the river. They moved a mile downriver, not knowing whether their brigadier was dead or alive. The next morning, five men set out to swim across the Chindwin. One drowned, but the others were found by British patrols. They reported the new location of the group, and again boats were gathered.

Wingate went out to the river with the rescuers. As they waited for the party's signal, Wingate spotted Japanese moving along the east bank. A minute later he saw the signal. The British replied, and got into the boats. As they were crossing the river, the Japanese opened up with mortar fire. However, the boats disappeared into the dark shadows of the east bank; in a few minutes they were full, and making their way back across the Chindwin. The Japanese mortar fire was again inaccurate—a blessing for the Chindit expedition.

Of the almost 3,000 men who had crossed the Chindwin with Wingate in February, only 2,180 had reached India four months later. The last to arrive was Major Ken Gilkes and men from his No. 8 Column. They had marched over a thousand miles north into Kunming China, and were flown back to India by the Americans. The first had been the men from No. 4 Column, of which Wingate had had no word of since February. They could not disengage from the Japanese, and rather than fall back towards the main force of Chindits, the column dispersed into small groups as they had been trained, falling back towards India.

Over 700 men had been killed or declared missing during the Chindit expedition. Of the survivors, more than that number would die within the following year as a result of the disease and debilitation they had suffered in Burma. This was a high price to pay in trained manpower for what the raid had physically accomplished. It had been an expensive failure as a military operation. What damage Wingate's Chindits had done to the Japanese railroads was quickly repaired. The number of Japanese killed was nowhere near as many as the number of Chindits who died, and a fraction of those who would never be fit for active service again.

What the Chindits did accomplish was to win a tremendous psychological victory. After a litany of defeat at the hands of the Japanese, Wingate had proved that the average Briton could defeat his savage enemy. Hungry for any good news, the British press trumpeted Wingate's achievement, comparing him to Clive, Lawrence and Hannibal. Wingate was recalled to England, where he was interviewed by Prime Minister Churchill as soon as he had gotten off the boat.

The Prime Minister was leaving for Quebec that night to take part in the Quadrant Conference. He persuaded Wingate to come with him, despite the fact that the officer was still tired, bearded, in his jungle bush jacket, and had not even let his wife know that he had arrived in Britain. The problem was solved when a Ministry of Defense auto delivered her to the dock from which Churchill's ship was departing. The Prime Minister prevailed upon Wingate to shave, bathe and keep his uniform clean during the conference.

One of the main items on the Quadrant agenda was the reopening of the land route to China. To this end, Roosevelt offered another revolutionary military unit, the newly raised 1st Air Commando Group. This was a composite group of a squadron each of bombers, fighters and transport planes. In addition, the group also had a massive number of gliders, light aircraft and even the first four helicopters produced for combat operations. They were led by someone who had all the leadership and charisma of Wingate himself. This was Colonel Philip Cochran, a distinguished fighter pilot who was already famous as an inspiration for the comic strip "Terry and the Pirates."

The pair hit it off, and Wingate was sent back to Burma. There, the 77th Indian Infantry Brigade served as a nucleus for the 3rd Indian Division. This unit would be a gigantic formation of five long-range penetration brigades, all of which were trained by Wingate as Chindits. Two of the brigades would be commanded by veterans of the 1st expedition. Brigadier Fergusson would command 16 LRP Brigade and Brigadier Calvert (with the new nickname of "Mad Mike") would command Wingate's old 77th Brigade.

The American airmen of the 1st Air Commandos, with the elite attached 27th Troop Carrier Squadron, fell under the messianic spell of Wingate. They loved his can-do spirit, which was in contrast to the condescending attitudes of the British officers they had dealt with while in India. Likewise, the Chindits loved the brash American airmen, who would do anything to help Wingate's men. To this day there is never a reunion of either the Chindits or the Air Commandos without representation from both groups.

Wingate's new plan for entering Burma was not a raid but an invasion. Operation "Thursday" would have four brigades of Chindits airlifted into northern Burma. They would be met by Fergusson's Brigade and Merrill's Marauders traveling overland with the objective of seizing of the Japanese stronghold of Myitkyina and opening the land route to China. Not only would Air Commando gliders bring the Chindits in over hundreds of miles of jungle, the unit's C-47s would supply them. The worst, most morale-shattering aspect of the first Chindit expedition was the fact that the sick and wounded had to be left behind to the mercies of the Japanese. Now the L-1 and L-4 light aircraft of the 1st Air Commandos could land on the most rudimentary airstrip and evacuate them.

The gliders started flying into the airheads during the night of March 5, 1944. Within four days 9,000 troops, 1,360 mules, and 250 tons of supplies were deep inside Burma. There would be hard fighting in the days ahead as reinforcements were flown in to hold the airfields and to release the Chindits for mobile operations.

During the night of March 24, Wingate hitched a ride back to his headquarters in India on an air commando B-25. It flew into one of the frequent thunderstorms over Assam, and crashed in flames. Since the remains of Wingate and his adjutant were badly charred, and could not be separated from those of the five American aircrew, they were all buried in a common grave in Arlington National Cemetery. Orde Wingate's body remains there despite official requests that were made by both Emperor Haile Selassie and David Ben Gurion to have him moved to a final resting place in their respective nations.

14

Second Schweinfurt

BY W. RAYMOND WOOD

One of the most devastating raids in history, and certainly the one with the most far-reaching consequences, was launched against Americans in the early morning hours of December 7, 1941. After the more than 300 raiding aircraft had returned to their carrier bases with the Japanese Combined Fleet, Americans at the stricken naval base at Pearl Harbor morbidly assessed their thousands of casualties and row of destroyed or sunken battleships. By the time the counting of its dead was completed, the United States had joined the global conflict. And having been brought into the war by an air raid, America would respond with an air capability of its own that would eventually become the most devastating in history.

But the growth of American airpower was gradual, and achieved in the teeth of fierce resistance from Germany and Japan. At first, in fact, American success was achieved more through courage than strength. Four months after Pearl Harbor, U.S. bombers led by Jimmy Doolittle were over Tokyo, on a one-way mission, simply to deliver to the Japanese a deadly message about what was to come.

In Europe, American planes arrived steadily throughout 1942 to augment the efforts of the RAF. By the spring of that year, British Bomber Command had created massive air fleets to attack the German homeland and had already launched its first "thousand bomber raid." The British, however, had suffered prohibitive casualties during daylight operations, and had taken to bombing at night. In the dark, they found that their bombers were unable to hit specific industrial installations, so they chose instead to direct their aircraft at the largest targets available: the centers of German cities.

When the American Eighth Air Force arrived on English soil, its first battle was with Bomber Command, resisting British efforts to subordinate the U.S. effort into its own nocturnal program. The Americans had their own ideas about how and where to hit Germany; they liked the daytime, when they could destroy anything they chose in precision attacks. And the Americans, not surprisingly, liked to attack industry.

Winston Churchill, bomber chief Air Marshal Sir Arthur Harris and other British leaders who had already been engaged in war with Germany for two years, were impatient with the neophytes from across the Atlantic and made a number of good arguments. Not only the British but the Germans themselves, during the last stages of the Battle of Britain, had switched to nighttime bombing because casualties were otherwise unsustainable. During the day, moreover, cloud cover often rendered bomber crews as blind in daylight operations as they would be after dark, and, further, clouds provided shelter for surprise attacks by fighters. The British also claimed that their targeting of cities was the quickest and cheapest means of bringing Germany to its knees. "Bomber" Harris derided smaller, industrial objectives as "panacea" targets.

General Ira Eaker of the Eighth Air Force, backed by Army Chief of Staff George C. Marshall and, ultimately, President Roosevelt, was nevertheless able to preserve U.S. operational independence from its more experienced ally. And the Americans thought they had a weapon that would prove them correct: an aircraft called the "Flying Fortress."

The Boeing B-17 was a fast, four-engine bomber armed with ten to twelve (depending on the model) machine guns that could spray deadly fire in any direction. Flying in tight "box" formations, the B-17 gunners could create fighter death traps in which an enemy pilot daring to enter the formation would find himself the target of 100 intersecting streams of .50-caliber bullets. Newly developed Norden and Sperry bombsights also gave U.S. bombardiers the ability to hit precision targets on the ground from heights of up to 25,000 feet. The "Flying Fortress," in addition to its great size, was also exceptionally sturdy and designed to fly on as little as one engine.

Americans were so confident in the B-17 and their other "heavy," the B-24, in fact, that at the beginning of its involvement in World War II, the United States lagged behind other nations in the development of fighter planes. When the first U.S. pilots arrived in England and were introduced to the Spitfire, many of them marveled at the British aircraft's grace and performance—far superior to the P-39 Aircobras and P-40 Tomahawks with which they had trained. In the months to come, the Americans would also become intimately acquainted with other high-performance fighter aircraft: the Messerschmitt Bf-109 and the Focke Wulf 190.

To German fighter pilots, the arrival of the B-17 in Europe was a challenge that prompted much discussion—as well as evolutions in both weaponry and tactics. For one thing, machine guns were not enough to bring a Flying Fortress down; cannon and rockets were needed as standard armament. The well-tested fighter tactic of approaching a bomber from below and behind was also no longer applicable. The fast speed of the B-17, as well as its armament, combined with that of other bombers in a formation, made the rear approach an agonizing crawl through a wall of fire. Instead, the Germans switched to making head-on attacks, three planes at a time on a bomber. The nose was the least heavily armored part of a B-17, and also the most lightly armed. Fighters attacking from the front could pick their target and then, benefiting from at least 550mph closing speed, fire and peel off before the Americans could react; and a hit on the cockpit would probably mean a downed bomber.

The first European American bombing mission in B-17s took place on August 17, 1942, and consisted of 12 planes attacking train yards outside Rouen, France. Other small missions followed through the rest of the year, generally aimed at U-boat facilities on the French coast. The Allies still lacked a long-range fighter that could accompany the bombers deep into Germany, but the Americans were gaining both confidence and strength. Although the main Allied effort had been redirected to North Africa and the Mediterranean theater in 1943, by the new year there were 500 U.S. "heavies" in England, that number increasing by the month.

At this time, the Luftwaffe fighter arm, led by Adolf Galland, was at its peak strength, and had behind it an extensive record of past victories to fuel the hunter instincts of its pilots. The German nightfighers were still evolving in their struggle to cope with British attacks on civilian population centers in the dark; but in the daytime, the Luftwaffe was experienced and ready, should the Americans attempt to practice their own concept of bomber war over German soil.

The American idea was that instead of employing their strategic air force as a blunt cudgel against anonymous civilians (although they would indeed do this later on a massive scale, most famously at Dresden and Hiroshima) they would instead identify the "nerve centers" of Germany's war industry. The Nazi military machine depended on surprisingly few elements—fuel, ore, machine parts—without which it would grind to a halt. And in the summer of 1943, the American braintrust discovered one group of small items that the Nazi state needed in order to survive: ball bearings.

These small machine components were essential to every phase of the German war industry, from tanks to planes to submarines. Furthermore, a full forty-two percent of German ball bearings were produced in the city of Schweinfurt. Because the Germans were thought to have almost no reserve supplies, the destruction of Schweinfurt's factories would be a catastrophic setback for Germany; a successful mission clearly would have shortened the war. The only problem was that Schweinfurt lay some 250 miles inside the German border.

On August 17, 1943, over 220 U.S. bombers ascended from their runways in the south of England, headed for Schweinfurt. Another 150 flew for Regensburg, where Messerschmitt factories were located, and thence to North Africa on the war's first "shuttle" mission. It was a clear day and Luftwaffe radar stations, from the Atlantic Coast to the Elbe River, tracked the bombers' progress. Fighters took to the air. By the time the raid was over, 36 bombers of the Schweinfurt contingent had been lost; 24 of the Regensburg. It was, up to that time, the worst disaster suffered by the U.S. Army Air Force in Europe.

For five weeks, deep-incursion daylight missions were suspended while the USAAF replaced its losses. As had happened in the past, the northwestern German city of Emden, located on the North Sea, suffered the brunt of the lull since it was a low-risk target. Meanwhile, Nazi Germany's Armaments Minister, Albert Speer, recognized what the Americans were up to and began dispersing the ball bearing production facilities. Nevertheless, the Americans, despite their losses, were determined to hit Schweinfurt again.

In England, in the early morning hours of October 14, 1943, a teletype message was issued by Brigadier General Frederick L.

Anderson to all the Eighth Air Force bomb groups assigned to participate in the forthcoming raid. It read: "This air operation today is the
most important air operation yet conducted in this war. The target
must be destroyed. It is of vital importance to the enemy. Your friends
and comrades that have been lost and that will be lost today are
depending on you. Their sacrifice must not be in vain. Good luck,
good shooting, and good bombing."

This was a heady incentive indeed for the men of the Eighth Air
Force. As U.S. lieutenant Elmer Bendiner later wrote, "An ordinary
young man is not often told that what he and a few others may do
between breakfast and dinner will change the course of world history." The mission to come would not, in fact, fulfill its desired
promise; but it would go down in history. To some it would become
known as Second Schweinfurt. To others, it has always been known as
"Black Thursday."

It was a large raid, consisting of three divisions of bombers. The
First Air Division, comprised of Fortresses in nine bomb groups, was
the spearhead, scheduled to reach Europe ten minutes before the Third
Division. Sixty B-24 Liberators made up the Second Air Division,
which was to fly a more southerly route—meaning it would reach the
target after the Third Division. All three groups were vectored farther
south for the return flight, because the weather was expected to go
downhill, with low ceilings and visibility that would hopefully handicap the German fighters. Intelligence estimates indicated that 500
German fighters were thought to be stationed in central and northern
Germany, while 200 others lurked in the occupied countries between
Schweinfurt and England.

On October 14, crews were awakened at 4 A.M., briefed at 7:00,
and a few hours later a British Mosquito reconnaissance plane dispatched to the target radioed home, "All of central Germany is in the
clear." The weather over Germany was indeed perfectly clear, with
almost ideal visibility. Schweinfurt, an ancient town on the upper
reaches of the Main River in Franconia (northern Bavaria), had a population of about 65,000, some 17,000 of whom were employed in the
ball-bearings industry. The principal targets of the mission were the
Vereinigte Kugelleger Fabrik (VKF) and Kugelfischer AG (FAG).
Other small factories, including those of Fichtel und Sachs Werke,
stood in the suburbs. Remembering the raid of August 17, residents of

Schweinfurt recalled it had been equally clear that day. The term "Schweinfurt weather" had consequently become so common that the town's citizens were uneasy about the fine weather on October 14. When the air raid alarms began sounding that afternoon in the city, no one there was surprised.

On the other hand, given the time of year, the weather in England was terrible. The famous fog lay heavy across the land, and drafts of murky air stirred slowly. The aroma of horses and hay was carried on the slight wind, since most of the country around the air bases was farmland and the harvest was nearing completion.

The men of the Eighth Air Force, carrying their awkward parachute packs, answered the call to "Board up!" and clambered into the planes through the waist gunner's compartment. Just before 10:00 in the morning the engines of hundreds of bombers were ignited across East Anglia as they prepared for a 10:15 takeoff.

As the pilots completed their preflight routine and watched the control towers, the "Go!" flares arched across the sky. The grass behind the heavy bombers was flattened by the swirling propellers as the planes moved from their hardstands to the perimeter track, awaiting their turn on the runway. Check flaps and rudder, and then apply full power. Takeoff! Visibility was down to a quarter of a mile because of the fog, and the bombers took off at one-minute intervals on full instruments, instead of the usual 30-second gap between planes. Slowly the B-17s rose into the murk, flying at 140 mph and using the needles of their radio compasses to home in on the radio beacons that would guide them to their assembly points aloft. The formations broke into bright sunlight at 6,500 feet and began to assemble behind their lead planes. Because of the fog, it took two hours for the entire force to assemble.

The First Division was commanded by Colonel Budd J. Peaslee, who flew as co-pilot with the 92nd Bomb Group's operations officer, Captain James K. McLaughlin, in the commanding aircraft. He was to lead the 383 bombers that were now assembling over England.

Zero Hour: the planes had assembled in formation and now were ready to approach the Continent. There were clouds over England, but the North Sea itself was clear below 10,000 feet.

The B-24s of the Second Division had so much difficulty in assembling that they were pulled off the raid, and twenty-four of them were sent to bomb a secondary target on the coast of Holland.

Most of the bomb groups formed up and headed east, save for the B-24s. Some aircraft in the remaining divisions aborted due to engine trouble, real or imagined. The stress of flying daylight bombing missions unhinged a small minority of pilots; the astounding fact is how many pilots pushed on despite the fear they felt. A further complication was the fact that because of timing, the 305th Bomb Group failed to link up with the 40th Combat Wing, meaning that a third of the lead wing's bombers were absent. Despite this glitch, Peaslee pressed on, breaking radio silence to order the First Combat Wing, which was behind him, to take over the lead. Standing orders forbade any wing composed of only two groups from entering German airspace because it was felt that such weakness invited destruction. Peaslee stayed in command, but was now in the second wing of the bomber stream.

The aerial armada left the English coast and headed east at 12:30 P.M., and began to cross the North Sea. Luftwaffe day-fighter control could not discern the target, but was well aware of the approximate size and direction of the force. Accordingly, radar stations all along the Dutch coast began monitoring the progress of the bombers, and fighter bases throughout northwestern Europe went on alert. Soon German fighters took off to begin the harassment, and were waiting along the coast of Europe when the First Combat Wing appeared over Walcheren, the southwesternmost island in Holland, at 1:00 P.M. They would normally have crossed into a different time zone, as the clocks in Europe are an hour later than in England, but for the duration of the war clocks in England were set two hours ahead in "double summer time," ensuring that they would register the same time as Germany despite daylight saving and the ordinary one-hour difference.

Because of the confusion and difficulty they had experienced in forming once they broke clear of the fog, the bomb groups were getting all mixed up. The Fortieth Combat Wing had only two bomb groups instead of three, and therefore a correspondingly smaller number of planes and guns. Other combat wings formed with two high groups, contributing to the clutter and confusion in the air. A total of 291 aircraft nevertheless persisted toward the target.

As they intersected the coast of Holland at their cruising altitude of 25,000 feet, the bomber stream was flying below a layer of scattered cirrus clouds, and the green and brown fields of Europe passed under them as they crossed the Netherlands and approached Belgium north of Antwerp. The skies became absolutely clear at the German border, but the sun brought no warmth to the men five miles above the enemy's homeland. The temperature outside the bombers was four degrees below zero and, with the windchill, a sixty-below "hurricane" blew in through the open waist windows where the gunners simultaneously froze and sweated.

The first German fighters tried to scatter the American fighter cover: fifty P-47 Thunderbolts had joined the bombers over the North Sea. One P-47 of the 353rd Fighter Group was shot down before their dwindling fuel forced all the escorts to return to base. Two B-17s also fell over Belgium, despite the fighter cover. The Thunderbolts (which had arrived in the European theater at the end of 1942) did what they could, and turned for home over Aachen at 1:33 P.M. The First Division bore the brunt of the initial German attacks, while the Third was relatively unmolested until it was well within Germany.

As soon as the P-47s departed, the full fury of the German fighters fell on the attacking bombers, and continued unabated all the way to and from Schweinfurt. The comment by a bomb group commander regarding First Schweinfurt was to be as appropriate in October as it was on that earlier mission: "It was like lining up the cavalry, shooting your way in and then shooting your way out again."

"We had no trouble until the P-47s left," one pilot recalled. "Then all hell broke loose. Between the Rhine and the target our formations were attacked by at least 300 enemy aircraft. Rockets mounted under the wings of enemy aircraft fired into our tight defensive formation caused the highest rate of casualties. The crews described the scene as similar to a parachute invasion, there were so many crews bailing out." At one point as many as 150 parachutes were in the air, representing 15 planes shot down.

A continuous air battle raged across central Germany as packs of fighters raced again and again into the bomber stream. They were mostly Focke Wulf 190s and Messerschmitt Bf-109s, but these were also joined by Messerschmitt Bf-110 Destroyers, which lurked beyond range of the bombers' machine guns and shot rockets into the formations from behind. The clear blue sky was crisscrossed by tracers and the crimson trails of the air-to-air rockets; wherever a rocket exploded near a B-17, the plane usually smoked and then burned, before plunging straight down or exploding.

The Germans' eight-inch rockets were adapted from the widely used *Werfer Granate* infantry mortar made by Mauser, and were fired electrically. The twin-engine planes carried four of the deadly rockets, while some of the Focke Wulf 190s and Messerschmitt Bf-109s carried two. The launch tubes for these rockets were called "stovepipes" by the German pilots, and when they were fired they drove their 250-pound, spin-stabilized rounds at 1,000 feet per second, adding high velocity to the destructive power of the 90-pound warhead. The preferred method of attack was to stay just outside the 1,000-yard range of the bomber's machine guns, and launch the rockets (fused to detonate at a pre-determined range of 600 to 1,200 yards) into the bomber formations. While they were not aimed, beyond simply pointing the fighter at the bombers, the effect of their bursting among the tightly packed B-17s was to force even the most hardened pilot to pull away, breaking up the formations.

When they went off, the rockets produced huge, bursting explosions, two to four times as big as the ordinary 88 mm anti-aircraft shells. A red streak marked their path across the sky, and the black burst of detonation had a dirty-red center. A single rocket could destroy a bomber, as opposed to the 20 to 25 hits with a 20 mm cannon needed to fell a B-17. Even the legendarily tough Flying Fortress had cause to fear these unguided missiles of death.

James E. Harris, flying with Butler in the 367th, the low squadron of the 306th Bomb Group, reported that "The German fighters hit us as soon as our escort left. I had never seen so many enemy fighters and planes of various types: The single-engine FW-190s and Me-109s rolled through the Bomb Groups, twin-engine and even sea planes lay out beyond our range and lobbed rockets at us, as bombers dropped aerial bombs from above into our formations . . . The intercom was bedlam with crew members calling out fighter attacks. It got so bad that it didn't do any good to call them out." Harris's plane was shot down on the raid.

The Germans were indeed throwing up every available plane against the stream of Flying Fortresses, some pilots landing briefly to refuel and rearm and then return to the battle. Recent German defeats in the Mediterranean resulted in assaults on B-17s by aircraft that had once supported Rommel's Afrika Korps. Captain Charles Schoolfield, leader of the 306th Bomb Group, described an Me-110 in his after-action report: "Looked like desert camouflage, shooting rockets."

As the bombers droned on towards their target, the German fighters seemed to be everywhere, plunging through the formations, trying to sneak up from behind, or screaming in from dead ahead. Every eye in a given B-17 was peeled for the glint of sun on a canopy or the flicker of light on a propeller, and targets were called out in bewildering bursts on the intercom as well as the radio. But there was almost no point to the frantic warnings, as one transmission could have covered the situation: "Bogies all around. Pick a target and drop it." Being only human, the pilots and crewmen called out to each other anyway.

Ammunition was expended by the B-17s at such a rate that in many cases gunners were out of rounds before they got to Schweinfurt. The Germans enjoyed the advantages of both speed and superior numbers, and their pilots pressed home their attacks with a verve that was only matched by the relentless progress of the bomber stream. The most unnerving attacks were from head-on, a twinkling ahead showing that a fighter had opened fire and was now closing on the nose of a bomber at tremendous speed. The Germans hoped that such attacks would break up the tight formations, making individual planes easy prey for fighters, but the Eighth Air Force pilots remained steadfast. They plowed ahead despite their casualties, intent on reaching the ball bearing factories.

Attacks came from below, behind, ahead and above, as well as on both flanks. Gun turrets spun and their machine guns loosed a torrent of .50-caliber bullets at any likely target. Eight-inch rockets, fired from two-engine planes, including Heinkel 111s, tore through the sky and left black clouds when they ignited. B-17s were hit and blew up in mid-air, while others dropped slowly out of formation, their engines on fire and their interiors wrecked by the machine guns and cannon of German fighters. To add to the carnage, anti-aircraft batteries on the ground sent a hail of steel aloft to deal with the interloping bombers.

Colonel Peaslee later wrote: "As the battle progressed, the hurts became more apparent. Throughout the formations bombers began to falter and fall behind or sink towards the earth far below. Some fuel tanks were ruptured and the burning, volatile gas force fed by the wind velocity ate through the soft metals of the aircraft structures, destroying aerodynamic qualities that held them aloft. From some of these bombers blossomed parachutes in quick succession, and on rare occasions all ten of the crewmen would float down towards the German POW camps. From others, when the machine died in violence, the crews were unable to reach the exits and became an indistinguishable part of the rubble when the bomber exploded on impact . . ."

At the outskirts of Schweinfurt the heaviest "flak" of the mission reached up to pelt the attackers. The bomber crews were almost glad to see it, because they assumed the fighters would pull away rather than fly through the bursts of their own guns. Instead, the fighters continued their savage attacks over the target itself, while the remaining bombers grouped together as best they could and dropped their 1,000-pound bombs in a rain of high explosives. Despite the best efforts of the Luftwaffe and German anti-aircraft gunners, the American bombers had reached into the heart of Germany to hit their target.

Having completed this stage of the mission, the survivors turned first south, and then west, to fly home. As they continued to battle the fighters, huge roiling billows of black smoke rose from Schweinfurt to indicate that destruction was heavy on the ground. But also visible were towering black plumes of smoke that marked the downward path of dozens of B-17s. These dismal pyres indicated the progress of the raid across Europe, and served as smouldering tombstones to mark where crews had gone down in pursuit of their objective. Until the target was reached, however, almost every Fortress that went down "was still pointing its nose at the target," as Captain John F. McCrary wrote.

Ten of the sixty bombers lost that day belonged to the 306th Bomb Group, more than half of the eighteen planes that took off that morning. Of the eighteen, three turned back early, seven dropped their bombs at the target and only five returned. The ten lost planes meant 100 men who would not get home, being killed or captured. One third of the men were killed, and two-thirds survived in POW camps. Severe as these losses were, the 305th Bomb Group exceeded them. Only two of its planes got home.

Still, 228 Flying Fortresses bombed Schweinfurt. But while the effort had been executed with great bravery, and the damage on the ground seemed extensive at the time, it transpired that the effect on the ball bearing industry was negligible.

The cost of the mission had been staggering for the Americans. All told, sixty bombers were shot from the skies and their crews lost; five more crash-landed or were abandoned upon returning to England; others that had managed to stagger back to England were so heavily damaged that they were reduced to scrap or cannibalized for spare parts. A dollar-amount can be set on the $21 million the Eighth Air Force spent in equipment that afternoon, but the cost of training those 600 men is hard to calculate, and the loss to their families defies numbers. One of the sixty planes lost that day was a B-17 called the *Wicked WAAC*, and one of the casualties was 2nd Lieutenant Elbert Stanley Wood of Nebraska.

"On this memorable day," Adolf Galland, commander of the Luftwaffe fighter forces said, "we managed to send up almost all the fighters and Destroyers which were available for the defense of the Reich, and in addition a part of the fighters of the 3rd Air Fleet, France. All together, 300 day fighters, 40 Destroyers, and some night-fighters took part in this air battle which for us was the most success-ful one of the year 1943. We were able to break up several bomber formations and to destroy them almost completely . . . On the German side about 35 fighters and Destroyers were lost." If Eighth Air Force intelligence underestimated the ferocity of German resistance, the B-17 gunners overestimated the 91 German fighters they claimed to have shot down that day. German newspapers realistically and properly claimed the battle a Luftwaffe victory, as opposed to General Henry H. "Hap" Arnold's announcement to the Allied press on October 18: "Now we have got Schweinfurt!"

Even as the raid was taking place, on American shores a new air-craft was being put through its final paces. After it arrived in Europe, it would go on to change the entire course of the daylight strategic bombing war. This plane, built by North American Aviation, was called the P-51 "Mustang." By the time the war was over, American bomber fleets would be roaming at will over Germany, protected by these long-range, high-performance fighters, destroying any target of their choice. The Mustang arrived in England on December 1, 1943, just six weeks after Second Schweinfurt.

15

Skorzeny at Budapest

BY STEPHEN TANNER

Although any political party that adopts a twisted cross for its emblem, holds regimented torchlight parades and favors black clothing for its officials might seem odd to the casual observer, such a party did, of course, take control of Germany in the 1930s. Nazi-run Germany subsequently achieved an impressive string of military successes, 1939–42, even if these were won by men in "field gray" rather than black leather. The Nazi leadership, however, had badly misjudged the determination of other world powers to challenge its existence, and the Germans and their allies found themselves overmatched. As victories came fewer and farther between—spaced, in fact, between monumental disasters—the Nazi state turned increasingly desperate and its methods became more byzantine.

In the two years after 1942, Germany was no longer able to wield irresistible power against its enemies, although it had not yet approached that Wagnerian "Götterdämmerung" phase of the war, when the Third Reich spiraled suicidally into a smoldering ruin. The history of the eventual death throes of the Nazi state would only be paralleled afterward by the inventions of science-fiction writers conjuring up mixtures of fanatic imagery and personal eccentricities. Yet from mid-1943 through 1944 there was a middling period: a gradual descent from triumph into the abyss, when Germany and its allies—the regular troops and citizens of the Axis powers—still had hopes of, if not victory, avoiding utter defeat for their homelands.

After the failure of their Kursk offensive in the middle of European Russia in July 1943, however, even the most optimistic of German military planners knew they were in trouble. At Kursk, for the first time without the help of "General Winter," the Red Army had beaten back the Wehrmacht, and its armor had held its own against elite German panzer divisions in history's largest tank battle. In Berlin, the horrifying reality set in that the Communist state was stronger than the Nazi one—and the Russians were coming.

Even as Erich von Manstein traded territory for cohesion in the east, on the Atlantic coast of Europe by mid-September 1943, the Germans were able to breathe a little easier. The Anglo-American coalition, which had steadily been building up strength for nearly two years, had declined to invade France. And the weather determined they would not be able to do so for another eight months: Germany still had time. But the Western Allies did invade Italy on September 8, following up their conquest of Sicily and their liquidation of Axis forces in North Africa the previous spring. The Allies' Mediterranean strategy, recommended by Churchill partly to create an access to southeastern Europe, might have played to Hitler's "grand strategy" (if not to Stalin's) by delaying the advent of a second front in France. But on the other hand it guaranteed that the Western armies would not be vulnerable to a massive counterstroke from the main forces of the Wehrmacht. The Allies could continue to grow stronger, building up men and materiel for the cross-Channel invasion that was sure to come. Meanwhile, huge bomber fleets streamed out of England both night and day to devastate, respectively, German cities and industry.

As for Germany's allies—Finland, Romania, Slovakia, Bulgaria and Hungary in the east; Italy in the west—by mid-1943, all had glimpsed the handwriting on the wall and were looking to desert the fascist cause. Perhaps because Anglo-American invasions provoke less fear in collective memory than the prospect of hordes of warriors streaming into Europe from central Asia, Italy was the first nation to desert the Axis coalition. After secret negotiations with the Americans and British, King Emmanuel II of Italy, in concert with Marshal Badoglio, took control of the government, arresting the fascist leader, Benito Mussolini, and spiriting him under heavy guard to locations unknown.

In Berlin, the defection of Italy resulted in a crucial debate. Field Marshal Erwin Rommel recommended that the Germans abandon the bulk of the huge peninsula and settle for defending the Alpine mountain passes in the north. Albert Kesselring, on the other hand, argued that the Italian "boot," crisscrossed by rivers and mountain ranges, comprised ideal defensive terrain, and that the Allies could be held off in the south for a year, maybe more. Adolf Hitler sided with Kesselring, who turned out to be correct, but in the meantime the German leader determined to rescue his friend Mussolini. For this task, he relied on an officer of the Waffen SS: a man named Otto Skorzeny.

The devolution of Germany from a respected, if feared, member of the international community into a unique historic example of martial, melodramatic excess is no better personified than by the figure of Skorzeny, the giant Nazi whose cheek sported a grotesque dueling scar. It's probable that only in such a gangster atmosphere as existed in the Hitler-run state could such a man, along with others like Goebbels, Goering and Himmler, have reached a pinnacle of national power at all. Nevertheless, Skorzeny was excellent at what he did—which was anything he was asked to do. As an unscrupulous, violent man at the service of the equally unscrupulous leader of the Nazi state, he would quite justifiably become known as "the most dangerous man in Europe."

Before the war Skorzeny had been an engineer in Vienna, and had joined the Nazi Party, he said, because of its motor racing clubs. When the war came he enlisted as a soldier of "Das Reich," the Waffen SS unit that would eventually be known as the 2nd SS Panzer Division. During the German attack on Yugoslavia in 1941, he recorded an exploit when he captured 52 enemy soldiers, but then, wounded outside Moscow after the invasion of Russia, and with gall bladder trouble besides, he was invalided home and assigned to a repair depot south of Berlin.

Skorzeny might have finished the war inconspicuously in the "motor pool," had not Hitler realized early in 1943 that the Abwehr, the German army intelligence service led by Admiral Wilhelm Canaris, was of dubious loyalty to the Nazi cause. The dictator needed to create a separate, and stronger, service from his loyal Shutzstaffel.

Within Himmler's SS, which grew exponentially throughout the war in direct proportion to Germany's misfortunes, Ernst Kaltenbrunner had succeeded the infamous (assassinated) Reinhard Heydrich as head of security. Under him, the elegant intellectual Walter Schellenburg was chief of SS foreign intelligence. To provide muscle for Schellenberg's spy organization, Otto Skorzeny was recruited by his fellow Austrian, Kaltenbrunner, to form a unit of clandestine commandos. The chain of command was Hitler–Himmler–Kaltenbrunner–Schellenburg–Skorzeny, but in practice Skorzeny would work for any one of his superiors separately, and without telling the others. At the same time, and particularly when he needed to call on Wehrmacht special forces and equipment, he consulted with the Abwehr, although he never let army intelligence know the fact that the Nazis knew they were disloyal.

Skorzeny was given free rein to recruit commandos from any part of the Third Reich, and established his training ground by a castle at Friedenthal, near Berlin. He trained his men for all types of operations, with various weapons, and schooled them in foreign languages. At one point, however, he admitted to Kaltenbrunner that he was unsure exactly what his men were supposed to be training for. "You're supposed to know," Kaltenbrunner replied, and Skorzeny turned to studying the known exploits of British commandos, as well as those of the Abwehr's own Brandenburg Battalion.

Skorzeny and his men spent the early summer of 1943 countering British operatives who were abetting the resistance movements in Occupied France and the Low Countries, curiously examining any secret British equipment they captured. But in July the overthrow and kidnapping of Mussolini presented Hitler with a personal crisis. With the Anglo-American invasion of Italy imminent, the Italian government was still professing loyalty to the Axis, but Hitler knew this to be a lie: "The House of Savoy will betray us this time," he said.

Six capable German officers were summoned to Berlin so that the Führer might choose a champion to attempt a rescue of Mussolini. His choice was the big SS man from Austria. Hitler told Skorzeny that only five people would know of his mission. "In no case," he said, "may the German embassy [in Rome] know of your order. Kesselring and his staff may not know of it under any circumstances." Strangely, as if both he and the SS major were overly concerned with having good reputations, Hitler later added, "If the mission is a failure, Skorzeny, I will find myself obliged to repudiate you in public and state that you acted on your own initiative."

The incredible rescue of Mussolini from an Italian mountaintop by Skorzeny's German commandos is a well-known story, and, if the protagonists had not been Nazis, might have been the subject of many a Hollywood movie. In short, Skorzeny arrived in Rome and proceeded to explore every lead that indicated where the fascist leader was being held. At one point, he found that Mussolini was imprisoned at Maddalena, an island off the Italian west coast near Sardinia. Skorzeny promptly drew up a plan for six E-boats, followed by minelayers packed with commandos, to race into the port, take the fort and rescue the dictator. But then he discovered Mussolini had been moved.

When he then learned that Mussolini was being held on the top of a remote mountain in the Apennine range, guarded by over a hundred soldiers, with many more holding the steep ascents, the only possible attempt at rescue would be a wild gamble. On September 23, 1943, German gliders nevertheless came silently swooping down on the peak of the 10,000-foot Gran Sasso. Their "landing area," essentially the backyard of a resort hotel in that rarified air, turned out to be strewn with rocks and boulders. One commando-filled glider was destroyed against the mountainside; two others turned back. Skorzeny's and six others all crash-landed. On seeing German commandos suddenly charging toward them across the grounds of their impregnable aerie, the Italian troops were stunned into paralysis. The SS leader dashed into the hotel and shot up a radio, then bounded up the stairs to the second floor where he found Mussolini, disheveled but unharmed. The Italian soldiers immediately capitulated to the Germans and Mussolini was freed.

The most hair-raising aspect of the rescue occurred when the issue of how to get Mussolini off the mountain arose. Motorized German Jaeger troops had secured the access routes to the mountain even as the glider attack was taking place, yet to spirit the dictator through a hostile population and difficult terrain against unknown opposition was not practical. A small, Fiesler-Storch scout plane had meanwhile landed atop the Gran Sasso. Mussolini was hustled aboard the light two-seater and the pilot prepared to take off. But then Skorzeny himself insisted on coming along.

The pilot protested that the Storch was not meant to carry such weight, and particularly not when taking off from a tiny, rock-strewn field atop a precipitous mountain. But the six-foot-four, 220-pound Skorzeny would not be overruled, and squeezed himself just behind Mussolini into the tiny scout plane prior to takeoff. The little aircraft whirled its propellor while commandos held onto its tail in order for it to get up power. Then it began waddling across the short field. Just before reaching the cliff, the Storch hit a rock with part of its landing gear and faltered; the plane skidded off the mountaintop and into a steep descent over the cliff. (The second thoughts of Skorzeny at this moment probably have never been accurately ascertained.) The pilot coolly devoted the dive off the cliff to the purpose of gaining speed, however, and somehow was able to pull out with full power before the Storch smashed into the valley floor with some twenty feet to spare. The plane then proceeded to Rome, skimming the treetops, since it was incapable of gaining altitude given the weight of its cargo.

Although the rescue of Mussolini did not result in the renewal of Italy's complete loyalty to the Axis, the Germans had demonstrated in spectacular fashion that they stood by their friends. And, in Skorzeny, Hitler had found an individual who could get things done.

The next world figure to find himself an objective of Otto Skorzeny was Marshal Henri Philippe Pétain, the leader of Vichy France. Hitler had received word that Pétain was plotting to collaborate with the "Free French" based in North Africa; Skorzeny thus traveled to Paris and thence to Vichy in the south. After requisitioning more troops from local German units, Skorzeny surrounded the town of Vichy and waited only for the codewords, "the wolf howls," before moving in. To his relief (since he admired the old French war hero), on December 20 the operation was called off. Germany elected to occupy all of France instead, since it needed to guard the French Mediterranean coast.

In early 1944, Skorzeny was assigned to capture or kill the powerful Yugoslavian guerrilla leader, Josef Broz Tito. Finding no help in Belgrade, he continued to Zagreb, nearer to Tito's known headquarters. German troops in Zagreb were shocked when Skorzeny pulled into town in his Mercedes, having incautiously driven through territory controlled by partisans. His efforts in Yugoslavia were wasted, however, when the local German commander sprang an attack of his own, Operation Knight's Move, against Tito's headquarters cave. Using Brandenburg special forces and Waffen SS paratroopers, the airborne assault succeeded in taking the hideout and also captured Tito's uniform, but the Yugoslavian leader had escaped by crawling out a back tunnel. Skorzeny disparaged the attempt: "All they found in his headquarters were two British officers, who [Tito] probably wanted to get rid of, and a brand new marshall's uniform." Skorzeny thought his commandos could have accomplished more.

An oddity of the Third Reich that, among other things, separated it from the Western powers during 1943–44 was that nearly every institution involved in the German war effort had plots afoot to kill the head of state. The phlegmatic Admiral Canaris at the Abwehr had been trying to kill Hitler for years, and by 1944 Schellenberg in SS intelligence also had schemes. The most determined group, however, was the German Army, which was bleeding itself white on half a dozen fronts in the service of a man many were convinced was leading Germany into a hopeless inferno.

On July 22, 1944, just six weeks after the Anglo-Americans had landed in France, Wehrmacht plotters were finally able to set off a bomb near Hitler, and then proceeded to take control of Germany. If Hitler were dead, the Army could seize power and fulfill its dream of calling off the war with the West and devoting all its resources to the East. The Waffen SS, of course, would have had a different view, as would Hermann Goering's Luftwaffe, which probably would have backed the SS. It has since been revealed that Propaganda Minister Goebbels received the news of the bomb with curious dispassion. As for the Army's ranking officers, the famous panzer leader Heinz Guderian experienced delays in dispatching two armored divisions to the Eastern Front on and around July 22, these having "paused" in the vicinity of Berlin.

In the first few hours after the bomb went off in East Prussia, no one knew whether Hitler was dead or alive, but a coup d'etat was obviously in progress in the capital. Hitler's loyalists frantically radioed Berlin for someone to pre-empt the insurgency. One man who happened to be on the spot was Otto Skorzeny.

The officer who had executed the assassination attempt, Claus von Stauffenberg, along with other plotters, arrived at the Reichs Ministry to take over the government. Skorzeny arrived minutes later, having been assured that Hitler was still alive, and with orders to crush the conspiracy. General of the Replacement Army Friedrich Fromm, in on the coup, exited the building for his home while everyone else was arriving; from there he would attempt to claim his innocence. Somewhere in that short sequence of time, Stauffenberg and his compatriots were seized, lined up against a wall and shot. Controversy still exists over who ordered their immediate deaths, although it was probably not Skorzeny, who could have better served the Führer if he had captured them alive. In any case, a thirty-six- hour period ensued when Skorzeny sat at the helm of the Third Reich, until the Nazis could re-establish their undisputed control.

In the weeks afterward, during Hitler's attempt to root out dissidents, Skorzeny became the messenger of his revenge. By the time the firing squads, suicides and hangings from meat hooks were finished, 200 German plotters (or suspects) were dead, and the Nazis were more firmly in control of Germany than before. It was at this time that the military salute, traditionally employed by the German as well as most Western armies, was replaced with the upraised right arm, signifying "Heil Hitler."

As for the war effort, by midsummer 1944 the Germans were bending, if not completely breaking, under the crushing power of their superior opponents. At Normandy, which was packed with Waffen SS divisions in the event that regular Wehrmacht units should prove "defeatist," the Anglo-Americans had broken out of their beachhead and were streaming across France. On June 22—the third anniversary of the German invasion—the Soviet Union had launched "Operation Bagration," a mammoth offensive that is sometimes simply known as "the destruction of Army Group Center."

On the Eastern Front in August 1944, another hammer-blow Russian offensive fell on the Romanian 3rd and 4th Armies. With his front destroyed, King Michael of Romania deposed the fascist strongman, Marshal Ion Antonescu, and on August 23 turned his country over to the Soviets. Luftwaffe planes then attacked Bucharest and Romania promptly declared war on Germany. In Berlin, with Ploesti already gone, the defection of the Romanian army to the other side was not considered a catastrophic event.

Bulgaria, never enthusiastic for the war against Russia, but which had allowed the Germans to use its territory as a staging ground, surrendered to the Soviets on September 9.

Finland, too, wanted out of the war, but the Finns were not to be caved in. Unlike Germany's other allies, Finland had not coveted additional territory but had only sought to reoccupy the land that Russia had bludgeoned out of her during the winter of 1939–40. When the Finns reached their previous border, north of Leningrad, they stopped.

In 1944, when it became obvious that the Germans would not be able to hold their position in Russia, the Finns sought peace terms from the Russians. As the situation worsened, however, Marshal Mannerheim fought on until German Army Group North had withdrawn beyond proximity to the Finnish border, thus evading the prospect the Finns would be made to turn on their ally. Then Finland agreed to another painful peace with Russia. Part of the agreement was that Finnish troops on the Murmansk front would seize the German 20th Mountain Army, which had been fighting in the far north, but in practice many Finnish ski troops essentially bid their old comrades farewell as the Germans withdrew to Norway.

Like Italy, Finland, Bulgaria and Romania before her, by the fall of 1944 it had become vividly clear that Hungary's decision to join the Axis had been a disastrous mistake. The Hungarians were considered to be excellent soldiers; however, their lack of native industrial plant had severely hampered their effort against the Russians. For modern equipment they relied on the Germans, who had scarcely enough to supply their own hard-pressed troops.

An entire Hungarian army had been annihilated in the eastern snow during the Stalingrad campaign, and now the Russians were at their doorstep. Nine Hungarian divisions still stood on the Eastern Front alongside the Wehrmacht, attempting to protect the Danube basin that had been, since the days of Attila, the traditional entryway into Central Europe from Asia. If Hungary fell, the Red Army would soon be in Vienna. But if Nazi Germany wanted to go down in flames, there was no reason for Hungary to be dragged down too. A respected nation, whose people had performed more than their share of defending Europe from the East down through the centuries, Hungary could arrange for itself a proper armistice, avoiding the "Götterdämmerung" that the Germans seemed to covet.

Admiral Miklos Horthy, the regent of Hungary, had first explored the idea of a rapprochement with the Anglo-Americans, who he hoped would soon be coming out of northern Italy. They had, unfortunately, only referred him to Stalin. By September 1944, Horthy had emissaries in the Kremlin suggesting a separate armistice. Hopefully, Hungary could declare itself neutral; at worst, it could turn against Germany, counting on Soviet troops to add muscle; the Wehrmacht could surely not maintain itself on another divided front and would have to withdraw from the country. The mercies of the Red Army could then be solicited in the postwar conciliation.

Whatever the arrangement, Horthy was aware that Hitler had eyes and ears in his country, and that the move to switch sides had to be made carefully and in secret. Horthy had always had a cordial relationship with Hitler, although not without some friction. Like most megalomaniacs, particularly those who take to calling themselves "the Führer," Hitler possessed a deep-seated craving for acceptance by members of social classes considered superior to his own. At one meeting between the two men, Horthy humiliated Hitler by claiming that he could not understand his German. To compensate, Horthy brought as his interpreter a corporal in the Hungarian Army who happened to sport a dark, toothbrush mustache. Nevertheless, Hitler valued Horthy who, after all, was a fascist and a dedicated ally, at least until the Russian steamroller appeared on his border.

Horthy's first son, Istvan, a respected fighter pilot in the East, had already died in battle. However, he had another, Niklas, on whom he now depended completely as his successor and heir. Together, Admiral Horthy hoped, he and "Niki" could steer their nation through the treacherous path between the main combatants. In the meantime, Budapest was packed with soldiers and guards.

On October 15, 1944, a German national, whose papers identified him as Dr. Solar Wolff, got out of his Mercedes in front of a row of apartment buildings in downtown Budapest. A Hungarian army truck was just feet away, and patrols canvassed the nearby park and alleyways. The big German, wearing a trench coat, opened up the hood of his vehicle and appeared to examine some trouble with the engine. At one point, the canvas of the nearby truck was momentarily lifted to reveal three Hungarian machine guns and a score of troops. Dr. Wolff continued to work on his engine.

Inside one of the apartment buildings was Niklas Horthy, who was meeting with representatives of the Yugoslavian leader Tito—another step in the defection of Hungary from the Axis. "Dr. Wolff," whose jagged facial scar that extended from his temple to his mouth would have identified him to the German High Command as Otto Skorzeny, calmly surveyed his hazardous surroundings. The several German MPs who were casually walking by on the street were really his commandos. In the nearby park, in plainclothes, others strolled, or sat on benches, pretending to read newspapers. "Dr. Wolff" had other men on the apartment floor above where the meeting was taking place. Thirty more commandos under his adjutant, Adrian von Folkersam, waited just around the corner.

The Germans had synchronized their watches and, five minutes after Niki's meeting with the Yugoslavians had begun, the "MPs" dashed to the door of the apartment building. The Hungarian truck, its canvas quickly raised, opened fire, dropping one of their number. Skorzeny revealed himself, only to come under a hail of bullets. His driver was shot in the leg and his Mercedes riddled with holes as he ducked behind its door. Both German commandos and Hungarian patrols came running from the park and the street became sprayed with crossfire.

Unexpectedly, Hungarian troops came charging down the steps of the building adjacent to the meeting house, where they had been placed as a secret back-up for Niki. Skorzeny jumped up, grabbed a grenade from the nearest German commando and heaved it at the doorway. The entranceway collapsed, preventing the troops from coming outside. Other German commandos tossed more grenades and the entrance was sealed. Folkersam's men had dashed into the square and the Germans now controlled the street.

The Germans in the apartment above Niklas Horthy's meeting room had meanwhile come down and seized the gathering. Skorzeny bounded up the stairs and ordered the Regent's stunned and bloodied son to be rolled in a carpet, the better to be kidnapped. He was carried down the stairs and thrust into a waiting car; the Germans moved off, even as a column of Hungarian soldiers—who had been hidden as reinforcements in an alleyway—came charging onto the scene.

At a bridge on the way to the airfield, Skorzeny and his men ran into a column of heavily armed Hungarian troops. Skorzeny leaped from his vehicle and admonished their commander. In the best tradition of "they went thataway!" he convinced the soldiers that the crisis was still going on back near the park. The Hungarians followed the orders of the authoritative German and moved off in the opposite direction. Skorzeny's commandos then reached the airfield and bundled their capture aboard a waiting plane for transport to Vienna.

The street battle, dubbed by the Germans "Operation Mickey Mouse," ended with the successful seizure of the Hungarian regent's son, to be held hostage for his father's geopolitical compliance with the Nazis. Admiral Horthy, however, though distraught, refused to be bent. Out on the front, Stalin, mindful that Hungary just needed one more push to capitulate, had launched the Red Army in an offensive against Debrecen. Hitler, just as concerned with keeping his ally in the fold, reinforced the Axis line with three panzer divisions. After a four-day tank battle, these fought back the Russians and destroyed three of their corps. The Russians were temporarily checked.

In Budapest, Horthy stuck to his convictions—Hungary needed to switch sides. In Berlin, the Nazis realized that their instinctive gangster tactics had not given them enough leverage. They needed to take Horthy himself. Skorzeny would have to follow up "Mickey Mouse" with "Operation Panzerfaust."

The Germans had already been pouring strength into the country, including 40 tanks into the capital. A gigantic, 650mm mortar, previously used in the reduction of the Warsaw uprising, was transported to Budapest. Skorzeny disdained the use of such blunt force. Being Viennese, he in fact held great respect for the Hungarians, and was convinced that their regular soldiers had no desire to surrender to Russia. He also preferred to achieve his goals, when possible, without gunfire. "Psychologically, it is of course easier to fire when attacking," he wrote. "[But] just one shot fired by the attackers is enough to awaken the self-preservation instinct of those being attacked, and they will automatically fire back." He continued, " I found an effective, proven means of preventing my soldiers from firing: namely to go in first and not fire myself."

Admiral Horthy made his headquarters in a huge complex over-looking the Danube called the Burgberg (Castle Hill). In its center was a castle known as the "Citadel," surrounded by steep ascents and 3,000 elite Hungarian troops. Fortified machine-gun emplacements were built into the approaches and the Hungarian garrison was armed with every manner of light and heavy weapons. Berlin had called for an airborne assault, but Skorzeny knew that such an attempt could only result in a massacre. He would instead attack the Citadel directly, and personally.

On October 15, Admiral Horthy broadcast over the radio that the Hungarian 1st and 2nd Armies were no longer allies of the Germans. He instructed their commanders to surrender to the Soviets, thus placing the entire German Army Group South in jeopardy of being cut off and destroyed.

Hitler's man in Budapest no longer needed to disguise himself as "Dr. Wolff." At the foot of the Burgberg, a column of gray military vehicles waited patiently for the order to advance. The troops seated within the vehicles sat quietly or smoked, while their leader walked purposefully up and down the column, exchanging final words with his subordinate officers. Anyone looking at his face would have tried not to fixate on the result of the saber slash he had suffered during his undefeated dueling career at the University of Vienna.

Skorzeny now had at his disposal the 2nd SS Cavalry Division, "Maria-Theresia," whose 8,000 Hungarians of German descent surrounded the general area of the Burgberg, facing outward. Closer in, he had been given 700 officer cadets from the military academy at Wiener-Neustadt. These would attack from the south, blow up the iron fence surrounding the castle grounds and pin down the Hungarian defenders in the gardens. From the Danube River, a battalion of Waffen SS paratroopers would break through a tunnel that led to the Hungarian War Ministry atop the Burgberg. Commandos would attack from the west, over the wall and against the front of the palace.

Skorzeny's own group would advance through the Vienna gate, on the road that led directly to the Citadel. He had two companies of Panthers plus a battalion of Goliaths, the little remote-controlled "tanks" that were ideal for revealing mines in the path of an advancing column. Nevertheless, Skorzeny ordered that the devices, as well as the Panthers, would follow him—Skorzeny himself would lead the approach to the Citadel.

The road was known to have been mined, and this was Skorzeny's greatest fear when, at six o'clock in the morning on October 16, the column began its advance. He stood upright in his command car at the head of the column. With him in the car was Folkersam and six other men, "chums from the Gran Sasso" as he described them. The commandos were armed with Sturmgewehr 44 assault rifles, grenades and panzerfausts (anti-tank weapons, not unlike bazookas). But their weapons were kept hidden beneath the sides of the car.

The column started up the winding road to the castle. On this occasion they disdained the element of surprise in timing, if not technique. One gunshot, one explosion—meaning the superior strength of the Hungarian garrison surrounding their head of state was resisting—would certainly culminate in the deaths of Skorzeny and most of his men.

So far no mines. Skorzeny came to the first roadblock: machine guns aimed at his vehicle. The German commandos nervously twitched their weapons, ready for combat. Skorzeny, however, simply saluted the defenders as his car crashed through the barricade. The Hungarians just stared. Closer to the castle, Hungarian strength increased, but the soldiers stood from their weapons and emplacements and stoically watched as Skorzeny stood in his car, progressing steadily upward to the castle itself.

At this point, Skorzeny's column could easily have been taken under rear and flanking fire from the Hungarians. Then, just when the Germans were at their most vulnerable, two explosions were heard. It was the SS battalion blasting its way through the tunnel to the War Ministry. "Drive faster!" Skorzeny ordered. The column thundered up the road to the Citadel. As it reached the fortress square it was met by three Hungarian tanks, their guns aimed at Skorzeny. The Germans screeched to a halt.

It was a tense moment while the commandos fingered their panzerfausts; but then the barrels of one of the Hungarian tanks moved. It raised its gun, indicating it would not fire. Behind the tanks, however, was a three-foot stone wall. Skorzeny instructed his driver to pull off to the right and then waved his arm at a Panther behind him. The German tank crashed through the stone wall, creating an opening for the other vehicles to pass through. The Germans then came face to face with six Hungarian anti-tank guns that protected the Citadel, and that held the intruders in their sights. But by now, two more Panthers had broken into the courtyard and had deployed. These guns, too, held their fire.

Skorzeny and his men were now dashing across the courtyard to the steps of the Citadel. A Hungarian officer aimed his weapon but Folkersam knocked it out of his hand. Skorzeny mounted the red carpet-covered marble staircase of the palace. In a hallway a Hungarian soldier had pushed a table against a window and was firing a machine gun into the courtyard. One of the commandos came up behind him, grabbed the gun and threw it out the window. Firing was heard from outside; not yet a full-scale battle.

Skorzeny burst through an ornate doorway to encounter a Hungarian general sitting behind a large desk. "Are you the fortress commandant?" Skorzeny demanded.

"Yes, but . . ."

"I request that you surrender the fortress immediately. There is fighting, can you hear it? Do you want to be responsible for bloodshed among allies?"

At that moment, a commando rushed in to announce: "Courtyard and main entrances, radio station and War Ministry are occupied. Request further orders."

Skorzeny and the commandant organized pairs of officers—one German, one Hungarian—to canvass the Burgberg and announce to all the troops that hostilities had ended. Commandos, from windows in the palace, also flung off a few panzerfaust shots in the general direction of the garden, as a signal to Hungarians still resisting there. The Germans lost four men killed, twelve wounded, the Hungarians three dead and fifteen wounded, in an operation that could well have cost hundreds more.

Admiral Horthy, it turned out, had vacated the premises minutes before Skorzeny's approach, turning himself in to General Karl Pfeffer-Wildenbruch, the ranking SS commander in Budapest. Since Otto Skorzeny had approached the seat of Hungarian government standing upright at the head of his "delegation," Horthy was not necessarily to be considered a traitor to the Axis cause. He was nevertheless provided with accommodations in a well-guarded castle within the borders of Germany, while a more ardent fascist, Ferenc Szálasi, was installed as Hungarian head of state. Horthy's radio signal to the 1st and 2nd Armies to capitulate had been jammed by the Germans (although some senior officers did defect to the Russians).

Prior to installation of the new government, Skorzeny was able to enjoy two days as master of the Citadel, sleeping in Emperor Franz Josef's old bed and bathing in his tub. At one point, Archduke Josef of Hapsburg, who would have been an emperor had that system been continued after the Great War, came up to Skorzeny and complimented him. "I heard that you are a Viennese, and that didn't surprise me. I said to myself: only someone from Vienna could pull off a stunt like that. Magnificent! Daring! I am happy to meet you. It's wonderful!"

The Archduke asked about the safety of his prize horses, kept in the royal stables. Skorzeny assured him: "Everything is as it was before." The Archduke tried to give him one of the horses, but Skorzeny demurred, informing the elderly man, "I command motorized troops."

Then it was back to the war.

For the Germans' last great counteroffensive against the Western powers, Hitler considered that they needed every advantage they could think of. To this end, at the Battle of the Bulge, Skorzeny dispatched about a hundred English-speaking commandos, dressed in American uniforms, behind the lines to sow chaos behind the front and to seize a bridge across the Meuse River. It is still not clear how the operation became interpreted as a plot to kill Eisenhower, but the Supreme Allied Commander was nevertheless held under massive guard throughout the duration of the battle. Skorzeny, who led a panzer brigade during the offensive, fighting primarily in the vicinity of Malmédy, later denied any intention to assassinate the Allied commander-in-chief.

By early 1945, Germany's military situation had become so precarious that, if it hanged by a thread, that thread has remained invisible to several generations of historians since. The continued resistance of Nazi Germany is paralleled only by civilizations in the pre-modern age, which fought on until the enemy was literally swinging swords over their heads. On the road to Germany's obliteration, Skorzeny had the bad luck of seeing the chief of the SS, Himmler, being named commander of the Army of the Oder on the Eastern Front. Himmler had no idea of how to command an army group, but in his hour of desperation at least had the power to call on his subordinate, Skorzeny, to command a bridgehead on the Oder River against the advancing Russian juggernaut.

The commando leader, although probably considering this task a waste of his skills, nevertheless performed well. In fact, his influence with the Wehrmacht and the High Command procured him more troops and equipment than other officers might have been able to muster, and he held his position against the Russians. Fortunately, Himmler was soon replaced by a legitimate Army commander who sanctioned, quite sensibly, the abandonment of what could have been Skorzeny's forlorn hope.

During the last months of the war, Hitler, driven underground, continued obsessing over his map tables, moving individual artillery pieces as well as entire divisions that no longer existed except on his map. Most other Nazi leaders, however, were more concerned with preparing for the postwar era. Skorzeny, as a "can-do" individual, was generally employed to spirit Nazi wealth out of Germany to safe havens in Switzerland, Spain, Portugal and other locales. In some cases he buried gold, jewels and currency in farmyards in Bavaria.

At the front, oblivious to the frantic scrambling of the Nazi leadership, German troops still fought on against the Russians. In the west, Wehrmacht resistance continued not so much from fear of Anglo-American occupation as to protect the backs of the men in the east, and to allow more time for refugees to escape.

In his underground bunker, meanwhile, Hitler must have wondered as to the fate of Skorzeny, unaware that his henchman was fully preocupied with getting cash out of the country. At the end, one of Hitler's last actions was one he had shrunk from throughout the war, or perhaps simply had procrastinated: he finally married his girlfriend, the former Eva Braun. The next day the newlyweds took cyanide; the Red Army had been a day or so away from being able to attend the ceremony.

After the war, the Allies tried Otto Skorzeny at Nuremberg and, although unable to gain a conviction, held him as a prisoner anyway as part of their de-Nazification program. In 1947 he "escaped" and went into business in Franco's Spain. While pursuing his engineering enterprise, he performed a leading role in ODESSA, the organization devoted to providing safety for ex-members of the SS. He also continued to be a confidant of notable world figures, including Juan Peron and his wife Evita, of Argentina, and Gamul Abdul Nasser of Egypt.

As for Hungary, the Red Army had been fifty miles from Budapest when Skorzeny toppled the government. Budapest held out until February 15, 1945, and was finally taken after a desperate, last-ditch defense by both German and Hungarian troops. The Burgberg, with all its historic treasures and artifacts (and Archduke Josef's horses), was destroyed by Russian aircraft and artillery. By that time, Nazi Germany itself was ten weeks away from annihilation. Although the distinction may be of questionable value, of all Germany's allies in the Second World War, Hungary was the only one to go down fighting.

16

Belgian Paratroopers in the Congo

BY ROBERT E. KROTT

As the sun rose over the trackless African jungle, heralding a new day for the inhabitants of Stanleyville in the former Belgian Congo, five United States Air Force C-130 Lockheed Hercules transport planes chased the dawn, not dropping their airspeed until just over the city's outskirts. The four-engined turboprops flew over Sabena airport at a height of 700 feet—combat parachute jump altitude. As the lead aircraft approached Sabena's main runway its navigator, First Lieutenant John Coble, yelled "Green light!" into the plane's intercom. Captain Robot Kitchen, the co-pilot, reached down to the control panel near his right armrest and as the paratroopers' jump warning lights switched from standby-red to go-green it was exactly 0600 hours, November 24, 1964. Operation Dragon Rouge, the meticulously planned airborne assault of the Belgian ParaCommando Regiment on Stanleyville was a "Go."

Lying dead center in Africa halfway between the Atlantic and Indian Oceans and halfway between Cairo and Capetown, Stanleyville (present-day Kisangani, in Zaire) is surrounded by the Congo's impenetrable jungle, a great tropical wilderness rivaled only by the Amazon rain forest. Beautiful, wild and mysterious, the jungle can also be a foreboding place where witchcraft, cannibalism and ritual murder is practiced. The sheer immensity of this Conradesque primeval jungle evokes awe mixed with a visceral dread—a sense that danger that could come suddenly from either nature or man.

On August 5, 1964, the "Simbas" of the Popular Army of Liberation under their self-commissioned "General," Nicolas Olenga, seized Stanleyville, the third largest city in the Democratic Republic of the Congo. Named after the Swahili word for "lion," the Simbas were a phenomenon that could only have transpired in the Congo. Witch doctors put the Simba recruits through an elaborate ceremony to confer powerful *dawa* (medicine) on them. The dawa would make them impervious to bullets; they only had to wave palm fronds and chant *"mai, mai"* (water, water) and look straight ahead while their enemies' bullets dissolved. Of course if a Simba was killed it was because he failed to follow the witch doctor's instructions. The troops were dressed in motley, although they donned any bits and pieces of military apparel that came their way. Their headgear consisted of everything from cowbay hats and abandoned UN helmets to lampshades. When the Simbas entered Stanleyville, General Olenga, a former postal clerk, appeared wearing the uniform of a Belgian lieutenant-general and carrying a dress sword.

Although without military training, the Simbas proved effective against the soldiers of the *Armee Nationale Congolese* (ANC), little more than a uniformed rabble, who themselves believed in sorcery. The ANC ran in terror from Simbas preceded by their palm frond-waving tribal witch doctors and their powerful dawa. (In truth, the ANC had few good marksmen; many soldiers believed it was the noise of their weapons which killed and so fired their guns with their eyes closed.) Despite the ad hoc nature of their rebellion, however, by fall 1964 the Simbas had seized much of the Congo, terrorizing and slaughtering thousands of Congolese as well as dozens of white missionaries, priests, nuns and expatriates.

Dragon Rouge's objective was to free the 1,600 foreign hostages being held by the Simbas in Stanleyville. These consisted of 500 Belgians, 700 other Europeans and about 400 Indians and Pakistanis. Also present were thirty Americans, primarily missionaries, but also including U.S. Consul Michael Hoyt and four other members of the U.S. diplomatic mission. Despite offers of evacuation by both the Belgian and American consulates when the Simba threat first appeared, the majority of these people had chosen to remain in Stanleyville. Hoyt and the other American diplomats were soon thrown into the Central Prison while other whites were held communally in hotels. The Simbas began making anti-American statements on Radio Stanleyville and issuing threats to execute their hostages if the United States did not withdraw all support from the Congo.

The Congo was no stranger to trouble. Shortly after independence in 1960 there was widespread political unrest leading to the Katanga secession of Moise Tshombe with the aid of white mercenaries under the command of a South African, Major Michael "Mad Mike" Hoare. Even as the Katanga secession chaos of 1960–63 took place, Antoine Gizenga had created a short-lived secessionist regime in Stanleyville in 1961. Still another rebellion, led by Pierre Mulele broke out in the Kwilu area near Kitwit in January 1964. Many of the whites in the Congo had survived these previous rebellions, and felt they could weather future storms as well. But later in 1964 a new rebellion, aided by the Communist bloc, began in the eastern territories of the Congo and rapidly spread. The Westerners simply weren't prepared for the ferocity of the Simbas.

At first the Simbas, whose political arm, founded by Christophe Gbenye, was called the "National Liberation Council" (CNL), numbered only a few hundred men poorly armed with spears and machetes and clad in rags and animal skins. Accompanied by their witch doctors, however, who claimed the dawa that could turn enemy bullets into water, the rebels moved from Albertville (present day Kalemie) on Lake Tanganyika with numbers soon swelling to thousands as they seized town after town without opposition. The ANC troops usually fled without firing a shot, abandoning their guns to the rebels. Although the Simbas were found to possess some Soviet and Chinese arms, the bulk of their men became supplied with weapons captured from the ANC.

On July 22, 1964, the provincial capital of Kindu, 250 miles south of Stanleyville, fell to the rebels. Despite the airlift of ANC troops to bolster the city's defenses, the Simbas captured Stanleyville. Many of the ANC surrendered, only to be tortured and then executed. The Simbas pushed on north and west of Stanleyville, eventually penetrating as far west as Lisala on the Congo River.

A week before the city's fall, Gaston Emile Soumialot, who was the Simbas' Minister of Defense, had issued a radio communique to the whites of Stanleyville: "You have nothing to fear from us. Stay here and continue to work. We need you. You will be happier under our government than under the Leopoldville regime. Nothing will happen to you if you remain in Stanleyville."

At first the whites were treated well by the Simbas but the situation soon deteriorated as the Simbas began to suffer their first setbacks at the hands of Hoare's "Wild Geese" mercenaries and their Katangese gendarmes. In Stanleyville, the Simbas began addressing the whites as *mateka*. Swahili for "butter," this word also meant "meat."

Americans were especially hated by Simbas. In fact, while they had been fighting mercenaries who were largely South African, Rhodesian, French and Belgian, the Simbas were convinced they were in combat against Americans, and their leaders blamed their setbacks on U.S. troops and aircraft. (General Olenga later claimed an atomic bomb had gone off in the Congo, and the Simbas were constantly on the lookout for an American submarine reported to be sneaking up the Congo River.) Soumialot issued a warning that any Americans captured were subject to execution and that he could also not guarantee the safety of UN officials.

The rebels under Generals Gbenye, Soumialot, and Olenga seized all the Westerners within their newly won territory as hostages. When confronted by Red Cross delegates, both Gbenye and Soumialot said they did not know what the International Committee of the Red Cross was. When told about the Geneva Conventions (particularly the ban on holding people as hostages) they said they had not heard about that either. In any case, they added, they did not consider themselves bound by the Geneva rules which, they scoffed, were "written by whites." At that time four of Gbenye's children were at school, in Geneva.

Behind Stanleyville's post office, in a picturesque park, was the Patrice Lumumba monument. On the morning of August 16, while Olenga was in Bukavu, one of his deputies ordered about a dozen Congolese prisoners to the monument. The Simbas then staged executions for the benefit of the gathered mob. Sometimes the victims were riddled with bullets. Other times they were hacked and chopped to bits with spears and machetes. Ritualistic cannibalism—the eating of certain body parts such as the heart or liver to acquire the power or virtues of the victim—was an old custom in this region of the Congo. Sylvere Bondeweke, a Stanleyville politician, was one of those killed at the Lumumba Monument. Bondeweke had to die, but Bondeweke had "power." The mob cut his liver from his body while he still breathed. Pieces were snatched up and devoured in order to gain Bondeweke's his strength. Over the next five days about 120 Congolese were slaughtered at this monument almost as some type of bizarre offering to Lumumba.

A Belgian, born in Stanleyville, met an old Congolese man (in Swahili, *mzee*) while the executions at the Lumumba Monument were still going on. The elder was downcast. "What's wrong, mzee?" asked the Belgian. "Life is hard, life is hard," lamented the old man. "They killed the mayor at the monument this afternoon and I cut a very nice piece from his back because I wanted some meat. Then I took it home. But when I left it near the fire to get some salt a dog grabbed it and now I have no meat to eat. I am hungry."

Throughout the Simba occupation, people were thrown off the Tshopo River bridge to the crocodiles below. As many as two thousand may have died in this way. At Kindu at least 800 were killed, many burned alive, at the Lumumba monument there. (It was in Kindu, on November 16, 1961, that 13 Italian airmen assigned to the UN forces were abducted and dismembered by rebellious ANC troops. Most of their bodies were actually sold in the Kindu market.) In Paulis, two to four thousand were slaughtered. Anyone who was an "intellectual"—a member of the old establishment—was singled out.

If not for the viciousness of the rapes, murder, mutilations and looting, the ignorance of the Simbas (especially the *jeunesse*, the young Simbas no more than 12- or 13-years-old and particularly given to butchery) would have been comical. In one home in Stanleyville, a group of Simbas came across a grand piano and were certain that it was a radio transmitter. None of them had ever seen a piano before. One European in Paulis wore a hearing aid which the Simbas believed was a radio transmitter used to call in American bombers, so they executed him. A Belgian in Boende often tapped his gold tooth nervously. The Simbas thought this must be another method of sending radio messages so he, too, was executed.

As the months passed without any military challenge to the Simbas, whites, as well as Congolese, increasingly became victims of atrocity. Throughout the Simba-held Congo, women, children and even nuns had been gang-raped before being killed. Something had to be done or the 1,600 white hostages in Stanleyville would surely meet with a dire fate.

Moise Tshombe, who had been called back from exile at the height of the crisis, in July 1964, to serve as the Congo's Prime

Minster, asked the United States for help. President Lyndon B. Johnson replied by dispatching Joint Task Force Leo. JTF Leo came about as part of the US Strike Command's (STRICOM) Operations Plan 515, generated during Mulele's rebellion. Several months later, General Paul D. Adams of STRICOM dusted off OPLAN 515 and revised it, adding a platoon of airborne infantry. The STRICOM task force of three Tactical Air Command C-130s and a C-130 "Talking Bird" communications aircraft were from the 464th Troop Carrier Wing at Pope Air Force Base, North Carolina. The aircraft were accompanied by a platoon of paratroopers from the 82nd Airborne Division. Total strength of JTF Leo numbered about 28 officers and 98 men. The U.S. Ambassador to the Congo had requested a battalion of troops—he got a platoon. Furthermore, this unit was required to secure approval from the White House before firing—or returning fire—at an enemy. (Shades of restrictions to come *after* 1964!)

A message intercepted from Simba General Olenga in Kindu to his commander, Gbenye, in Stanleyville on October 7 (the same day that Mike Hoare's 5 Commando took Uvira) said. "I give you an official order. If NATO aircraft bomb and kill Congolese civilian population,

The Congo in 1964, showing the advance of the Simbas from the east.

please kill one foreigner for each Congolese in your region . . . If no bombing, please treat foreigners as honored guests in accordance with Bantu custom. Give them food and drink."

While a flurry of negotiations involving American, Belgian, and African diplomats ensued, various military solutions were investigated. Both Belgium and the United States began planning for military intervention. Special Forces A-Teams were readied for a possible rescue mission; "Operation Flag Pole," a plan to rescue the consulate staff in Stanleyville with an ad hoc force of Marine embassy guards and former military types in Leopoldville, was called off (according to one American Embassy official, "It was a half-assed operation anyway."); Joint Task Force Leo was sent into the Congo to evacuate whites not being held by the Simbas; and an assault by paratroopers of the 82nd Airborne Division was planned.

As the situation in the Congo worsened, William Brubeck, a representative of the Congo Working Group, met Etienne Davignon, the *Chef de Cabinet* for Belgian Foreign Minister Paul Henri Spaak, at a reception on November 8 and the American was told by the Belgian, "If you get us the aircraft we can do it with a battalion."

It sounded like a good idea.

While there is no record of Spaak's offer or of the meeting at President Lyndon lohnson's ranch where Secretary Rusk and National Security Advisor McGeorge Bundy presented the proposal to President Johnson, Lieutenant General David Burchinal, Director of the Joint Staff, called the US European Command Headquarters (USEUCOM) in Paris the very next day (Armistice Day evening in Paris) with orders sending three planners to Brussels. Brigadier General Russell Dougherty, USEUCOM Deputy Operations Officer, was called away from the annual Marine Corps Ball to leave for Brussels. He was joined there by Lieutenant Colonel J.L. Gray (USAF Headquarters, Europe) and an airborne operations expert, Captain B.F. Brashears (US 8th Airborne Division). With them was Colonel J.E. Dunn, carrying verbal instructions from the Joint Chiefs of Staff: "Keep it small. Don't let it turn into a John Wayne operation; no pearl-handled pistols or wagons so loaded the mules can't pull them."

The planners worked throughout the 12th and 13th of November. The plan, USEUCOM OPLAN 319/64 (Operation Dragon Rouge) was a mere 22 pages long. Dragon Rouge would be a three-phased operation with twelve US C-130E aircraft flying into Kleine Brogal, Belgium to load 545 paratroopers of the *Regiment du Para-Commando,* eight armored jeeps and 12 AS-24 motor-tricycdes and then fly 4,134 nautical miles to Ascension Island, stopping to refuel enroute at Moron de la Frontera, Spain. Phase II was a 2,405-mile flight to their staging area at Kamina Air Base in Katanga. Phase III would be the 550-mile flight to Stanleyville (escorted by B-26s flown by CIA-hired Cuban pilots) and the parachute assault.

The parachute assault was also broken down into three phases: 1) Five US C-130s would drop 320 Belgian paratroopers from the Paracommando Regiment (the 1st Battalion arid Regimental Headquarters) onto the golf course just northeast of the Stanleyville airport. 2) Then, following the assault force by a half hour, two more C-130s loaded with the armored jeeps would land at Stanleyville's airport. 3) Thirty minutes later the remaining five C-130s carrying supplies and one company of paratroopers from the 2nd Battalion would land or drop their men and supplies as needed. It was a simple plan. No John Wayne. No pearl-handled pistols.

On November 15, Brigadier General Robert D. Forman, commander of the US Air Force's 322nd Air Division, was ordered to prepare to airlift a Belgian force to the Congo. Colonel Burgess Gradwell, commander of Detachment One of the 322nd at Evreux, France would have take charge of the Dragon Rouge airlift. The aircraft were assembled at Evreux Fauville Air Force Base, France, and at 1740 on the 17th the fourteen C-130s departed for Kleine Brogel, Belgium.

Flying in the first aircraft were Colonel Gradwell; Captain Donald R. Strobaugh, the 5th Aerial Port Squadron (APORON) combat control team commander; and Sergeant Robert J. Dias, a 5th APORON radio specialist. When they reached Kleine Brogel the Belgian paratroopers were loaded. Their commander, Colonel Charles Laurent, age 51, had led the paratroops who seized Leopoldville's airport in 1960 from rebellious ANC troops and had jumped into Stanleyville Airport in 1950 on a training exercise. Laurent's sergeant-major had also jumped into Stanleyville. While the sergeant-major had over 3,000 jumps and Laurent over 300, this was not true of the young troopers in the regiment. Most were 18- to 20-year-old conscripts. The paratroopers of the 1st Battalion had the most time in service: 10 months!

The C-130s departed at 2240 enroute to their refueling stop at

Moron Air Base before continuing on to Ascension. When the C-130s landed to refuel, an American major left one of the aircraft to supervise the refueling and to ensure the planes remained closed to observation. But someone opened a door for fresh air and one of the ground crew looked inside. Sitting inside were the Belgian paratroopers wearing their red berets. The surprised airman reported to the major: "Sir, who are those guys in red hats?"

"What guys in red hats?" replied the major.

"Them guys in red hats!"

"*What* guys in red hats?"

The airman looked again at the planeload of young Belgian paratroopers wearing their distinctive red berets. The major closed the door saying. 'There's nobody in there wearing red hats!"

Arriving at 1310 on November 18, the paratroopers and the C-130 crews waited on Ascension for three days. During this time Captain Strobaugh taught twenty-one Belgian jumpmasters proper C-130 parachute drop procedures and, along with Sergeant Dias, taught the radiomen how to use the PRC-41 and PRC-47 radios used for ground-to-air communications. Because of a shortage of Belgian jumpmasters, Captain Strobaugh would jumpmaster the ninth plane, Chalk-9, but was under strict orders not to jump himself. On November 20 the unit commanders were briefed on the assault plan. Strobaugh relayed the details to Washington and at 1800 the men of Dragon Rouge were alerted. They didn't have to wait long; a half-hour later the "Go" order was given. At 1935 the first aircraft lifted off for Kamina.

After nine hours in the air, the Dragon Rouge force landed at Kamina, which was still obscured by fog at dawn. At Kamina the commander of JTF Leo, and now the overall commander of Operation Dragon Rouge, Colonel Clayton Isaacson, gave the force a briefing. The paratroopers were forced to wait at Kamina while Belgian and U.S. diplomats continued negotiating for the release of the hostages.

Meanwhile Lima One, a column of trucks and armored cars carrying ANC troops and "Mad Mike" Hoare's 5 Commando, and led by Commandant Albert Liegeois, 48, a Belgian Army officer, had been fighting its way from Kongolo, 470 miles away, through the jungle to Stanleyville. The hundred or so English speaking mercenaries and the 150 Katangese soldiers of Lima One rescued many Europeans and discovered the bodies of others who'd been executed and mutilated. Along the way they killed thousands of Simbas.

At Kindu, Lima One was reinforced by another group of mercenaries bringing their strength to 120 white soldiers and was joined by another motorized column, Lima Two. Lima Two consisted of 50 French-speaking white mercenaries and 300 ANC troops under Commandant Robert Lamouline. Like Liegeois, Lamouline was on loan from the Belgian Army. Both units were under the overall command of Colonel Frederic Van de Waele. Van de Waele was a former Force Publique officer with over 25 years service in the Congo. It was hoped that this force would link up with Laurent's paratroops. Van de Waele had already convinced Laurent to delay the parachute assault for a day so that his force could strike Stanleyville at the same time. Some American embassy officials disagreed. It would be politically safer for Laurent to rescue the hostages on Monday, the 23rd of November, a day before the mercenaries arrived. Others agreed with Van de Waele. As it was, the assault was delayed a day because of negotiations with the Simbas. The jump was scheduled for Tuesday, the 24th. If the paratroops couldn't jump because of weather, then Van de Waele's force would have to rescue the hostages themselves. Without the surprise of the parachute assault it might be too late.

For many others it already was. On the 19th at Isangi, about 100 miles downstream from Stanleyville, the Simbas had killed two nuns (an American and a Belgian) and a Dutch priest. Many other nuns were beaten and raped.

On the 24th at 0130 local time, just three hours after receiving the code word "Punch" from the Joint Chiefs of Staff, the commander of JTF LFO took off from Kamina to rendezvous over Basoko with support aircraft and get a weather report for Stanleyville. The weather was good. Operation Dragon Rouge could proceed. The first five C-130s, designated Chalk-1 through Chalk-5, lifted off from Kamina at 0300 hours

At 0600 the first five transports dropped their 64 paratroopers apiece at twenty second intervals over the Stanleyville golf course. The paratroopers seized the airport against minimal resistance and cleared the runway of its obstructions—wrecked vehicles and 55-gallon fuel drums. Three Simba vehicles were caught attempting to flee the airfield. Two were destroyed by gunfire and the third captured. Inside the third vehicle the Belgian paratroopers found Simba leader Gbenye's passport, ID card, vaccination record, and a large sum of cash. Resistance was light and most of the Simbas fled into the bush.

The Cuban-piloted B-26s circled above, hungrily looking for targets. Two T-28 fighters shot up the roads surrounding the airport as the pilots—a Cuban and an American mercenary—watched the Red Berets spread out in a perimeter. So far the mission was a success and other than three paratroopers injured in the drop there were no friendly casualties. The telephone in the control tower, in use as Colonel Laurent's command post, rang at 0635 and someone in Stanleyville informed the paratroopers that the hostages were in Victoria Residence Hotel and in danger of being executed. Laurent sent the paratroopers into the city.

With the runways clear of obstructions and the airport perimeter secured, the other seven C-130s were ordered to land. Chalk-6 and Chalk-7 with their armored jeeps were to be followed by Chalks 8 and 9 carrying the company of paratroopers from the 2nd Battalion (these were troops with less than six-months service) and then Chalk-10 carrying the three-man motortricycles, Chalk-11, which carried ammunition, food, and medical supplies, followed by Chalk-12, the hospital plane.

While Chalk-7 with four of the armored jeeps landed at 0645, Chalk-6 with the other four was delayed an hour. Aboard Chalk-6, a liferaft had accidentally inflated and obstructed the plane's tail controls. The pilot, Richard V. Secord (later of Iran-Contra fame), saved the aircraft and was then able to deliver his vehicles to the paratroopers. Secord was subsequently awarded the Distinguished Flying Cross.

Using two armored jeeps, paratroopers of the 11th Para-Commando Company entered the outskirts of Stanleyville a mere three kilometers from the airport at 0740, meeting with light resistance from the Simbas at each street intersection. They were several blocks from the Victoria Residence and the hostages at 0750 when they heard shooting.

About 300 hostages, including 17 Americans, were being held in the Victoria Residence. When the Simbas heard the approaching C-130s they began ordering all the hostages out into the street. Nearly fifty were able to hide themselves in closets and under beds. Of the 250 or so hostages lined up in three columns outside the Victoria, about 100 were women and children. Only a dozen Simbas guarded them. While five or six had automatic rifles, the others carried spears and machetes. The Simba commander, "Colonel" Joseph Opepe, a former ANC officer, ordered them marched to the airport. Near the Congo Palace Hotel, just two blocks from the Victoria, Opepe ordered the column to halt. Firing could be heard.

Opepe ordered the hostages to sit down. Suddenly a group of Simbas rushed up and informed Opepe that Belgian paratroops had landed at the airport. Confusion reigned for the next few minutes. Then heavy gunfire was heard nearby, and getting closer. Without warning the Simbas began massacring the hostages.

For nearly five minutes the Simbas shot at the hostages, most of whom fled. Some were chased down and stabbed or hacked to death by the Simbas. US Consul Michael Hoyt and Vice-Consul David Grinwis escaped. Phyllis Rine, 25, a missionary from Ohio, was fatally wounded in the leg. Also killed in the shooting was Dr. Paul Carlson. Carlson, a missionary of the Evangelical Covenant Church, had been declared "a major in the American Army" and sentenced to death by a Simba "war council." Diplomats from around the world had pleaded for the missionary doctor's life. Carlson fled after the initial fusillade of Simba bullets only to be cut down as he tried to scale a wall which Charlie Davis, another American, had already cleared.

Finally, taken under fire by approaching Belgian paratroopers, the Simbas fled before they were able to kill more people. When the Red Berets arrived they found two little girls, five women and fifteen men dead or dying and forty other whites wounded (five would later die of their wounds). The priest with the 11th Company gave the last rites to the dead and dying. The survivors, including Consul Hoyt and four other consulate staff, were escorted to the airfield.

Laurent's men continued clearing Stanleyville, engaging the Simbas in sporadic firefights, and effecting the rescue of hostages wherever they found them. The paratroops suffered their first combat casualty at about 0900 when a paratrooper was wounded—shot in the back. Half an hour later a second para was wounded and at 1000 a third, who died two days later.

The first plane, loaded with 120 evacuees, lifted off for Leopold-ville at 0945, and at 1100 the four lead vehicles of mercenaries from Van de Waele's column entered Stanleyville. Hoare's mercenaries continued clearing Stanleyville while Laurent's paratroops concentrated on evacuating the civilians and securing the airport. There had been approximately 150 rebels west of Runway 28 and another 180 rebels in foxholes in the treeline parallel to the runway. Later that afternoon 150 Simbas unsuccessfully attacked the airport. In all, the Belgians repulsed five attacks. They would continue to take fire throughout the day and were engaged in several firefights that night. The sporadic, incoming small arms fire from the perimeter forced Strobaugh and his combat control team to orbit aircraft and change runway direction. Several of the C-130s boasted bulletholes. In the morning a Belgian airman was killed by a sniper.

Meanwhile, Van de Waele's force was having problems crossing the Congo River to the city's left bank, until Mike Hoare and his mercenaries finally forced the crossing on the 26th. By then, however, they were met with a grisly sight: the massacred bodies of 24 priests and 4 Spanish nuns. After herding the Catholics into a small cell, the Simbas had shot them and then cut their throats.

With the mercenaries and the ANC now in force in Stanleyville the *ratissage,* or "raking up," began. Unfortunately, the Congolese of Stanleyville would find very little difference between the Simbas and the ANC troops, who were now bent on pillage and revenge. The ANC often considered all civilians found in a formerly Simba-controlled area to be Simbas, or guilty of collaborating with the rebels, and treated them accordingly. Regarding the conduct of ANC troops in Stanleyville, Captain Stobaugh noted in his journal: 'The senseless butchery and mutilation of civilians and military prisoners is unbelievable. There was absolutely no difference between the everyday acts of the ANC and the well-publicized atrocities commited by the rebels. Only the uniforms were different. . . . I wonder if the people of the Congo might possibly find that the cure is as bad as the disease."

Over the next few days Van de Waele's command and the Belgian paratroops would find many more massacred whites. But during the Stanleyville operation a total of approximately 2,000 foreign nationals and 300 Congolese were evacuated on 80 aircraft sorties.

Immediately after Stanleyville, Operation Dragon Noir (Black Dragon), a similar airborne assault on the city of Paulis, took place. Because of the success of Dragon Rouge there was concern that the Simbas would massacre whites being held in other cities. In fact,

Hoare's 52nd Commando (a sub-unit of 5 Commando) was already moving on to Paulis after seizing Aketi on November 24th. At Aketi, the mercenaries had rescued 117 adult Europeans, along with 17 children, but they reported that whites held in Paulis were at risk. It was decided to launch a second airborne operation on the 26th against Paulis and Bunia and then move overland to Watsa.

Colonel Laurent, however, felt his troops were too tired and his force too small to accomplish simultaneous assaults and rescues. He would settle for hitting Paulis on the 26th with two companies of paratroops. The flight to Paulis took only 52 minutes, and at exactly 0600 the C-130s were over the airfield, which was shrouded with fog. The pilot of the lead aircraft overflew the dropzone and had to orbit. This alerted the Simbas. The four C-130s discharged their cargo. The last paratrooper out the door was shot in the chest. The other 255 "red berets" landed safely.

Because the 27 hostages massacred in Stanleyville were killed less than two hours after paratroopers on Chalk-1 had hit the ground, Laurent ordered his 11th Company to proceed immediately into Paulis while the 13th Company seized the airfield. Chalks 5, 6 and 7 began landing at 0640 hours. By 0315 almost 200 refugees had been gathered at the airport. They had a horrible story to tell. On the night of the 24th the Simbas had executed 21 Belgians and 1 American. After the Simbas tortured him for nearly an hour, Joseph Tucker, an Assembly of God missionary from Arkansas, was the first to die. The victims were then thrown into the Bomokandi River. Thirty-five others, including Tucker's wife and three children, were rescued. By the afternoon of the next day Operation Dragon Noir had rescued and evacuated 375 foreign nationals.

Despite the presence of other Europeans at Bunia, Watsa, and Wamba, Laurent's paratroopers were withdrawn from the Congo. Many debated the decision then (and since), as "Mad Mike" Hoare and his mercenaries remained behind, continuing to rescue hostages and recover the bodies of the Simbas' victims. Some were no doubt slaughtered in retaliation for the assault on Stanleyville. When Wamba was taken nearly a month later it was learned that 30 hostages had been killed during Operation Dragon Noir, and the total of white hostages killed after Dragon Rouge was estimated at 185.

17

The Israelis at Green Island

BY STEVEN HARTOV

In the Middle East, there is no such thing as a mild July. It was the summer of 1969, American soldiers were dying in record numbers in the jungles of Vietnam, the United States was preparing brave-hearted astronauts for their first lunar landing, and although the State of Israel's morale still surfed on the waves of its Six Day War victory, Israeli and Arab blood continued to flow freely in the sands of the Sinai Desert.

In June of 1967, threatened on all fronts by the combined forces of the Syrian, Jordanian and Egyptian armies, the Israelis had launched a lightning pre-emptive strike, their air force eliminating its Egyptian counterpart before the Soviet-supplied aircraft could be rolled out onto the tarmacs. Israeli armored forces had smashed through the Sinai and the Golan Heights, while Israel Defense Force paratroops recaptured Jerusalem from the Jordanian Legion in bitter house-to-house fighting. For a short time, it seemed that scores had been settled in the Levant, with Israel emerging as an unbeatable modern Sparta.

Yet it was not long before the Middle East returned to its eye-for-an-eye status quo. Resupplied by their Soviet masters, the Egyptians punished the Israeli forces occupying the Sinai with repeated artillery engagements across the Suez Canal. And although still unable and unwilling to engage the vaunted IDF in a pitched battle for the desert peninsula, the Egyptians encouraged and supplied various terror factions for strikes against the Israeli population. The IDF responded with its own artillery strikes, air raids and counter-insurgency efforts, and a full-scale though undeclared conflict called the War of Attrition was taking its toll on all sides.

It was to become the golden age of Israeli special operations.

While the new Jewish State had barely passed its twenty-first birthday, it had already been engaged in three all-out wars of survival. The War of Independence of 1948, the Sinai Campaign of 1956 and the Six Day War of 1967 were generously interspersed with major cross-border and counter-insurgency operations of a grand scale, and it seemed that the periods between declared conflicts were little more than rest and recreation furloughs for embattled veterans. From the inception of the Israel Defense Forces just after World War II, special operations units began to play significant roles, first as forward reconnaissance units of elite airborne and armor brigades, and later as independent outfits tasked directly by the general staff. The Israeli Army borrowed a page from the British Special Air Service and formed the ultra-secret commando outfit known as Sayeret Matkal, or General Staff Reconnaissance Unit. The Israeli Navy studied the methods of WWII Allied and Axis frogmen, forming its own SEAL unit encoded as Shayetet 13—Flotilla 13—more commonly known as Ha'commando Ha'yami, or the Naval Commandos.

Sayeret Matkal and the Naval Commandos were developed with decidedly different missions in mind. The Sayeret was a long-range reconnaissance unit, generally operating in small teams behind enemy lines, performing surveillance operations, targeting key enemy personnel for capture or elimination and raiding terrorist strongholds. The Naval Commandos played a more traditional role, silently entering enemy harbors to sabotage warships and naval installations, or striking key enemy ground forces when an attack from the sea seemed impossible and certainly improbable. However, both units drew their personnel from the cream of Israel's young recruit crop. Kibbutzniks, the collective farmers raised on hardships and cooperative social endeavors, volunteered for these outfits with enthusiasm, although the rigorous selection processes weeded out all but the most robust, intelligent, determined and stubborn applicants. Operators from either outfit had to ultimately be able to parachute, endure forced marches of incomparable distances and terrain, swim though rough and frigid seas with absurd loads of gear, and then fight like tigers in close-quarter battles.

Both during and after the Six Day War, Sayeret Matkal's reputation within the IDF soared as the unit conducted a series of daring long-range raids. Israel's national air carrier, El Al, was being repeatedly targeted by Palestinian terrorists for hijackings and blatant assaults. In response, in December 1968, Matkal spearheaded a heliborne raid into the heart of Beirut International Airport, blowing up thirteen aircraft of Middle East Airlines without sustaining a single casualty or causing any to the shocked civilians who witnessed the event.

In contrast, although certainly comprised of equally talented operators, the Naval Commandos suffered a dearth of luck. During the Six Day War, Commandos were launched from the Israeli submarine INS *Tanin* to attack the Egyptian port of Alexandria, but six of the men were captured in the unsuccessful raid. Despite the embarrassment, naval commanders insisted that the Commandos be given a second chance, and a series of ambushes and sabotage missions conducted "postwar" across the Suez Canal built up confidence in the sea raiders.

In June 1969, the Egyptian radar facility at Ras El-Adabiya was slated for assault by the IDF general staff. Fed up with Egyptian artillery attacks, the Israelis were planning a large-scale air force hammering of the Egyptian positions, but paving the way for such an effort demanded some pre-emptive blinding of Egyptian "eyes and ears." El-Adabiya was reportedly manned by up to fifty enemy soldiers and ten technicians, and initially the Naval Commandos were to serve as merely a ferrying service for a larger force of IDF paratroopers. However, the Shayettet CO at the time, later to become commander of the Navy, was Lieutenant Colonel Zeev Almog (Israeli navy personnel use army ranks), and he lobbied long and hard for his men to prove their mettle in a solo assault on the position.

On June 21, 1969, Operation Bulmus 5 was launched from the IDF base of Ras Misala in the Sinai. A commando force of twenty operators, including a five-man fire support team, made the initial crossing in Zodiacs, then managed a thirty-minute swim to the beach. Taken completely by surprise, the Egyptians lost 32 KIAs and 12 wounded, in addition to having their radar, power systems and signal equipment destroyed by satchel charges. The Israelis spent a total of twelve minutes on target and withdrew with three minor casualties.

Once again, Sayeret Matkal and Shayettet 13 seemed to be in a

dead heat in the competition for "best of the best." However, Israeli
successes seemed to bolster Egyptian resolve, rather than dampen it.
Israeli paratroops pulled off two deep penetration raids on June 29
and July 2, killing 13 Egyptians and taking 3 prisoners. In response,
Egyptian commandos raided an IDF tank refueling depot on July 9,
killing 8 Israelis, wounding 9, and taking one Israeli prisoner, whose
corpse was found the next morning in the nearby desert, having been
summarily executed. In celebration of their battlefield coup, Egyptian
artillery pounded Israeli positions all along the length of the Suez
Canal.

Israeli Minister of Defense Moshe Dayan, together with IDF Chief
Of Staff Chaim Bar Lev, decided that a "message" needed to be sent
to the enemy that no Egyptian position, no matter how well-fortified
or apparently impregnable, was out of reach for Israeli special ops
forces. In a meeting before the general staff, Colonel Zeev Almog once
again raised his proposal for just such a raid. It was to be against Al
Ah'dar, the Egyptian early warning station, otherwise known as Green
Island.

Green Island was a fortress built by British forces during World
War II. Located in the Gulf of Suez, just four kilometers south of the
city of Suez and the mouth of the Suez Canal, the structure sat atop an
eight-foot high seawall ringed by three wide rolls of razor concertina
wire. The island, which was 145 meters long and 50 meters wide,
served as base for a radar station, four 85mm anti-aircraft guns, two
37mm anti-aircraft guns and fourteen machine-gun positions. The
Egyptian garrison consisted of approximately 70 infantrymen and 12
As-Saiqa commandos.

Despite the healthy competition between the army and the navy,
Almog proposed a combined operation of Naval Commandos and
Sayeret Matkal. In a first-wave breaching assault, the Commandos
would approach underwater via Swimmer Delivery Vehicles, then
clear the perimeter for the second-wave Zodiac assault by a combined
Matkal and Commando force, while a third force set up diversion and
covering fire from a nearby rocky outcropping just south of the island.
The basic plan was approved, and Colonel Menachem Digli was slated to command the Matkal force, while Almog acted as overall CO for
what would be known as Operation Bulmus 6.

As with all Israeli special operations, preparation for the raid would consist of intensive examination of air reconnaissance photos, recon surveillance, repetitive "dress rehearsals," and frank and heated discussions among the officers and operators without regard to rank or veteran status. Practice raids on "models" forced modification of the original assault plans. The raid on Ras El-Adabiya had been performed after a surface swim, but the Green Island defenders were already extremely wary of such an attempt, leaving the Israelis no choice but an initial approach to be executed underwater. The Israelis were utilizing indigenously manufactured SDVs called "Pigs," but given the likelihood of an intense battle on Green Island and the possibility of having such a vehicle fall into enemy hands, the SDV option was declined. The Pigs would only be used to deliver the diversion and cover team to its remote rock.

Although the Naval Commandos had never before attempted a long-range dive with heavy assault gear, this option was selected after a series of tests were carried out underwater and in pressure chambers to determine the effectiveness of ammunition, grenades and RPG rockets after exposure to depths. A method of linking the divers was borrowed from the tactics of Italian frogmen, and two lengths of rope with handhold loops were devised for the first-wave assault team. Meanwhile, IDF army intelligence reviewed British Admiralty charts to determine wave and current strength around the island, while a recon observation post set up at Port Tewfik revealed that members of the Egyptian garrison repeatedly came down to the water through the perimeter fence on the northwestern side of the island. Apparently, there was some sort of breach in the defensive structure there.

In a general staff meeting held in "The Hole" below IDF headquarters in Tel Aviv, Moshe Dayan gave his green light for the mission, to be launched on the night of July 19. The one-eyed war hero, already an international celebrity with his bald head and buccaneer eye patch, emphasized that should the assault force suffer more than ten casualties, Bulmus 6 would be declared a failure and the raiders withdrawn. General Bar-Lev set out mission priorities: extraction of wounded, destruction of the positions, taking of prisoners.

On July 17, a truck convoy bearing the mission's heavy gear headed down through Sinai for Ras Sudar on the east bank of the Gulf of Suez. On the next day, the mission commanders and operators arrived, beginning a series of preparations and briefings. Laying out a mission timetable, a window of one hour—beginning at 01:30—was given for commencement of the actual assault to withdrawal from the island, the second wave moving in only upon signal from the first. Target priorities were designated: blowing the big 85mm anti-aircraft guns, then the main northern fortress building, and finally the radar and elint camp.

Lieutenant Dov Bar of the Naval Commandos was selected to command the first wave of four teams, each including two officers and three commandos. Lieutenant Elan Egozi's team was selected to breach the barbed-wire fences, if possible by locating the present breach, and if not by using a water-proofed Bangalore Torpedo. Dov Bar's team would then move in, utilizing ropes to climb a high structure and cover the rest of the first wave's assault. The third team, commanded by Lieutenant Gadi Krol, would cross a connecting bridge and take out an enemy observation and defensive tower—a precarious venture at best, considering that the machine gunners in the tower covered that very bridge. The fourth team, commanded by Lieutenant Amnon Sofer, would proceed to the north building and clear its interior. There was no hard intelligence at all on what awaited the Israelis inside that building.

At 1400, final briefings were held with the Chief of Staff, the CO of the air force, the OIC of all Sinai forces and other personnel of high rank. The attendance of so much brass unnerved the mission operators, who were further demoralized when Bar-Lev emphasized that, while the mission was important, it was not to be accomplished at any price. Israeli special ops personnel were used to functioning "at all costs," and the implication that here, on Green Island, the price might just be too high did nothing to relieve their tension. Lunch was served in a silent mess, operators from the Commandos and Matkal huddling separately, and when the teams finally boarded their transport trucks and headed for their point of departure at Ras-Misala, each man remained quietly secluded on his own, private island of dark thoughts.

At 2000 hours, the silent black waters of the Gulf of Suez lapped at the rubber prows of 12 Zodiac Mark Vs and "Pigs," and at 2030 hours the raiders slid away from the beach. The first-wave boats were commanded by naval officer Shaul Ziv, followed by the second-wave boats, commanded by Dani Avinon, ferrying the Matkal and second Commando teams, as well as the Command and Control team, including Zeev Almog himself. Also aboard was General Rafael "Raful" Eitan, Chief Paratroops and Infantry Officer, later to become IDF Chief of Staff.

The Naval Commando operators wore only their upper wet suit halves to defend against the cold waters of the Gulf. Over the wet suits and their bare legs, they wore quick-drying dacron fatigues, and canvas sneakers inside their swim fins. Load-bearing vests laden with ammunition, grenades, flashlights, ropes and first-aid kits were fastened down over their combat apparel, then came inflatable BC vests and, finally, oxygen rebreathers. Personal weapons included Uzi submachine guns or Kalachnikov assault rifles. There was no need to seal the weapons or ammunition, as everything had been successfully tested to a depth of six meters.

At the time, sure-fire range finding devices had not yet been developed for such operations, but General Eitan had devised a simple and reliable substitute. At a range of 900 meters from Green, "Raful" raised a homemade "fork" in his hand, the distant image of the fortress settling perfectly between the "prongs." The assault force had reached its jump-off point, and the first wave went into the water.

Dov Bar's Rolex showed 2300 hours when his Commando contingent of twenty men began their surface toward Green Island, the gulf waters cold but the swells calm, a mild 5 knot wind drifting across the waves. In optimum conditions, the range could easily be closed in an hour's swim, attaining the island at 1230, but experienced special ops commanders knew that things never, ever proceeded as planned. A large window had been left for the swim, and it was assumed that the 0130 assault time would surely be met.

Uzi Livnat and Yossi Zamir swam point on their backs, each trailing the long guide ropes linked to their harnesses, each operator gripping his loop and hauling 40 kilograms of assault gear. After an hour, Bar realized that the group was making little progress, and that the gulf's surface currents did not remotely resemble the intelligence assessment. Bar decided that progress might improve below the surface, and he opted for an earlier dive than planned.

It took another ten crucial minutes to set up the dive, but finally Bar was on point, leading with a compass, the team ropes linked to his harness. Livnat and Zamir followed Bar, working to keep their commander moving as freely as possible while the rest of the commandos tried to maintain their positions and buoyancy. Shepherding the team proved to be a hellish effort. The waters of the gulf were all too clear, a fierce moon reflecting off the surface. Bar wanted to keep his men at an undetectable depth, yet letting them sink below 4 meters might induce oxygen poisoning as well as corrupt their ordnance, and only the group's three officers possessed depth gauges. The commandos struggled to make progress against a stubborn current, while their monstrous loads threatened to drown them. It was a supreme test of the Naval Commandos' penchant for training under the most extreme conditions.

After some time, Dov Bar surfaced for a recon check—only to find that he was still 600 meters southeast of the target! The hour was 1230, and Zeev Almog and Raful were already trying to make contact by radio. Bar ignored the attempts of his commanders, fearing that if he reported the truth, the mission might be called off. More than anything else, "peer pressure" spurred the young lieutenant on, because he could not imagine returning to Flotilla 13's base of operations with his tail between his legs.

With hand signals, Bar raised his assault team to the surface and hissed at them, cursing in a harsh whisper that they were going to make it, come hell or high water. Counter to everything they had learned, they were going to remain on the surface and swim in close, only diving again at the last minute. His fellow officers tried to dissuade him from this tactic, insisting on an immediate dive to maintain surprise, but Bar knew that he would never make the timetable that way, and he hauled on the team ropes and kicked off again for Green.

Meanwhile, the diversion and cover team arrived at their position in their SDV Pig. A 4 x 4 meter bunker rose out of the water on a crust of rock and coral, and the team tied off their Pig on the south side of the outcropping and proceeded to set up their machine guns, ammo crates and a bazooka. The second wave boats, trailing the empty Zodiacs of the first wave, gathered around a distant buoy, tied off and waited, engines running, for some kind of signal from Dov Bar.

Bar and his men swam to within 150 meters of Green, in full view of the Egyptian machine-gun tower, before the lieutenant again gave the order to dive. The going was no easier than previously, but spurred by their progress, the men pressed on. Bar came up for a number of "sneak and peeks" to check the range, and at 15 meters from the seawall the gulf bottom began to slope, and the Commandos gratefully felt their flippers touch down on coral.

Bar signaled the men to dump their diving gear, and they doffed their oxygen rigs, tying them to the rope leads while still breathing off the regulators. Their swim fins went into special pouches on their LBEs, inflatable vests were left in place and dive masks were also retained, in case it would be necessary to make a hasty withdrawal before the second wave brought in the Zodiacs.

Still below the surface now, Bar cocked his weapon, while his men followed suit. He checked his watch: 0138. They were well past the outside limit for "commencement of hostilities," but there was no going back now. The OIC whistled inside his mouthpiece and the entire assault team doffed their regulators and silently breached the surface next to the retaining wall. Bar knew that Zeev Almog was probably frantically trying to make commo contact, but it was no time for a chat.

Bar hand-signaled his breach team and six men crawled along the retaining wall, which proved to be sloped at a helpful angle, rather than vertical. The Israelis rarely use camouflage paint of any kind, but on this night it was particularly unnecessary, as the Commandos were now covered in a thick, black oil slick. Bar pulled himself up onto a small cement platform, covering his men while the first barbed-wire fence was easily cut and the remainder of the assault teams huddled beneath a short bridge connecting the radar installation to the main structures of the fortress. The Egyptians were on their guard, sentries walking and talking above, while the Israelis held their breaths, still ragged with exertion.

The second row of concertina proved to be more troublesome—taut and threatening to twang like a banjo with each cut—and a sentry tossed a cigarette into the air, the sizzling butt landing within centimeters of Bar's teams. If the Egyptians discovered the Israelis before the attack commenced, the Commandos would be sitting ducks in the mud below the wall.

Bar looked around for an alternate route, and suddenly spotted a gap in the concertina wire nearby, a pile of trash forming a small delta below the gap. He remembered the recon reports—Egyptians coming down to the sea at regular intervals—and he "knocked" on his own head and pumped his arm, signaling the entire assault team to follow him toward the gap. One man doubled over to serve as a "ladder" and the commandos began to mount the seawall.

Without warning, a bright light flooded the center of the island. An Egyptian with a flashlight headed toward the wall, ramming the bolt back on his AK-47. Lieutenant Elan Egozi, just breaching the top of the wall, was sure that he and his men had been spotted, and he immediately opened up, blowing the Egyptian off his feet and momentarily shocking the members of his own team. Seconds later, an Egyptian grenade bounced off the wall and exploded, felling Egozi with a brace of shrapnel to his legs, but he yelled at his men to charge.

For a moment, the assault team was stunned into immobility, but Dov Bar quickly rallied them with shouts, and the attack began. Egyptian flares arched into the sky, and all at once the night attack turned into a "daylight" raid. From the coral rock just south of the island, the diversion and cover team opened up on the island with everything they had. The bazooka man, who had never used the rocket tube in action before, was knocked back into the water by the back blast, and a comrade hauled him back out and set him up for another shot.

At first, the Egyptian response was sporadic and inaccurate, and the Commandos used those precious seconds to scale the seawall and split into their designated teams. While half the attacking force moved to block and clear the southern "fort" section of the island, a four-man team fought their way across the bridge to the radar facility. At the very least, if the Israelis were beaten back off the island, they would destroy the radar. Two commandos battled a spitting machine-gun nest atop the radar bunker with grenades and submachine-gun fire, while a second pair cleared the interior of the bunker itself.

At this point, the Israelis began to discover that half their ordnance was useless. The weight of their gear had taken them to untested depths, and many of their grenades, both smoke and shrapnel, were duds.

Although the Israelis still carried the battle, the momentum of their attack driving them on, the Egyptians began to respond fiercely with counter-thrusts and accurate gunfire. There was no "fight or flee" option on Green Island. The Egyptians could not withdraw, and the raid quickly developed into a series of close-quarter gunfights. At every corner turned, the Naval Commandos faced garrison soldiers and As-Saiqa springing from doorways and bunkers; scores were quickly settled at point-blank range. Bar's men climbed the roof of the main building using one man's back as a stool. Lieutenant Ami Ayalon found himself facing two machine-gun posts, and although his and his men's grenades all failed, they took out the post with their personal weapons. Commando Israel Assaf realized that the Egyptians were hammering his comrades from the observation tower, and he single-handedly destroyed it, firing his Uzi and hurling grenades until one of his frags finally worked.

At this point, Dov Bar attempted to call in the second wave by radio, but was unable to raise them. However, by prearranged backup procedure, he fired a green flare to summon the Matkal commandos, then followed it with a red flare to signal the covering team to hold their fire. His men were now passing through the crossfire, and the cover team on their coral rock stood down and began remounting their Pig.

Laden with Matkal commandos and the Command and Control Team, the second wave of Zodiacs roared into the island. The first-wave Naval Commandos were almost out of ammunition, and the Matkal operators stormed over them, setting up a CP on the bunker roof with Zeev Almog and Menachem Digli commanding the operation from that point. Almog ran a quick check of his casualties: the Naval Commandos already had two dead and six wounded, and almost instantly a Matkal officer was killed by a bullet in the forehead as he led his team to take out one of the heavy gun emplacements.

Meanwhile, a battalion surgeon and a medic had arrived with the second wave and immediately began tending to the wounded, while Almog made radio contact with the rear, prepping helicopters and ambulances on shore for evacuation. The demolition teams moved in behind the main assault forces, placing almost two hundred pounds of explosives on the gun emplacements, headquarters and barracks.

Firefights raged across the island. Because the entire position was not much larger than a football field, there were no "long range" exchanges. It was hell in a very small space, both Israelis and Egyptians spinning to respond to each new threat, the roar of close-quarter gunfire, grenades and RPG rockets deafening them and suppressing their shouted orders. But with the arrival of the second wave and fresh assaults by the Matkal commandos, Egyptian fire began to slack off, and the Israelis witnessed a number of the enemy throwing themselves from bunker roofs into the sea in a last-resort effort to save themselves.

At 0225, Raful ordered a withdrawal, with the evacuation of casualties a priority. An extraction team had already recovered the first wave's diving gear, coxswains carefully loaded the Zodiacs with fixed numbers of men to maintain a head count, and the attacking force began to withdraw. However Almog, Digli, Bar and two other officers remained on the CP, holding off the surviving Egyptians while the Israelis searched for two missing men. The Egyptians called in artillery on their own position, destroying one of the Israeli Zodiacs, but at 0255 the satchel charges were set and the remaining Israelis piled into Zeev Almog's Zodiac and made for open water. Fifteen minutes later, as the Israeli commanders roared away across the Gulf of Suez, they turned back to watch as the charges erupted, illuminating the gulf waters in a flash of light and hurling rock, destroying what remained of the Green Island fortress.

For the Naval Commandos, the exfiltration from Green Island was no less a nightmare than the infiltration. The Egyptian forces in the gulf area had reacted to the raid like killer bees in a trampled nest, opening up with artillery all across the Israeli shore positions. The assault force's Zodiacs, now loaded with three dead Naval Commandos, three more dead Matkal operators, and eleven wounded, made for landings on a number of beachheads, only to be driven off by intensive barrages. Israeli tanks were summoned down to the shore to serve as "ambulances" for one flotilla of Zodiacs, while choppers were called in to search the gulf waters for the remaining boats.

One Zodiac, its hull peppered with shrapnel, was sinking fast, its engines useless. The craft only survived to make it 400 meters from the flaming wreck of Green Island when Lieutenant Amnon Sofer ordered it abandoned. He and his men went over the side with their weapons, commo and signal gear for chopper extraction, and a five-liter jerrycan of water. They sliced up the Zodiac with their diving knives and began the long swim toward the eastern shore. Because heavy radio traffic crammed the nets during the rescue efforts, Sofer and his team were in the water until 0500, when a chopper finally spotted them and pulled them from the gulf.

It was over. On that very morning, Israeli newspapers would laud the raid as Israel's Navarone, but many of the raiders were to remain in military hospitals for months to come, and all across the country IDF cemeteries would be crowded with tearful mourners, exhausted officers and proud politicians. Yet in the Middle East, it was just another blazing summer day.

The Green Island Raid was a watershed for Israeli special operations, although the mission still continues to evoke controversy within the IDF. Although 36 Egyptians had been killed in the raid and their entire facility was destroyed, the Israeli raiders sustained nearly fifty-percent casualties, and much heated discussion ensued as to whether the mission was properly evaluated, planned and executed. Yet rather than recoiling from further special operations endeavors, the IDF used Green Island as a model for improvement. Just after the raid, the Israeli Air Force engaged the Egyptians in a series of dogfights and bombing raids, causing over 300 enemy casualties. And, in the ensuing months of the War of Attrition, the Naval Commandos staged nearly 80 small unit raids across the Suez Canal. Sayeret Matkal also continued to build its international legacy, culminating in the most famous anti-terror rescue operation of modern times, when its operators served as the point men for Operation Thunderball, the Raid on Entebbe.

Although the Green Island Raid, in retrospect, might have been executed differently, there was no denying of the brave actions under fire of the raiders themselves. The Israel Defense Force rarely issues medals for valor, yet three of the Naval Commandos were awarded the highest such distinctions for their courage on Green Island, and Flotilla 13 took its rightful place among the very best of the world's special operations units.

18

The Americans at Son Tay

BY MICHAEL F. DILLEY

In 1970, the nine-year Vietnam War was at a turning point. Two years earlier, after U.S. military and intelligence organizations had assured President Lyndon B. Johnson that the enemy forces were weak and losing the war, the Communists launched a series of attacks known as the "Tet Offensive" that took everyone by surprise. Although the Communist forces were soundly beaten on the battlefield, they had achieved a major propaganda and media victory. Later that year, Richard M. Nixon was elected president, largely by promising to end the war. After almost 18 months of secret and not-so-secret talks to fulfill that promise, Nixon authorized U.S. military forces to conduct a limited incursion into Cambodia in an operation aimed at the Communist headquarters in the south, COSVN (the Central Office for South Vietnam). Public opinion, already hardened against the government, exploded in rage at this expansion of the war. Even though the physical location of COSVN was overrun, no major victory resulted from the Cambodian incursion. More emphasis was then placed on "Vietnamization"—turning the prosecution of the war over to the South Vietnamese themselves in order to permit the withdrawal of U.S. forces. Negotiations between the U.S. and North Vietnam continued, with both sides emphasizing, in different ways, the prisoner-of-war issue.

In and around the city of Hanoi, the capital of North Vietnam, were several holding stations and compounds that housed American pilots and air crew members as prisoners. These Americans had been shot down by anti-aircraft batteries or by surface-to-air missiles

(SAMs). As the Americans were brought into the prison system most of the "new guys" were taken to the isolated part of the old French prison, Hoa Lo, that they eventually called Heartbreak Hotel. Here they went through some initial interrogation and, when virtually all of them refused to give more than their name, rank, service number and date of birth, they were introduced to "the ropes."

Despite having signed the 1957 Geneva Accords on Treatment of Prisoners of War (a fact it denied throughout the war), North Vietnam made it a matter of government policy to brutalize and mistreat the Americans and, later, other Allied prisoners of war. This brutality was inflicted for even the slightest excuse. Initially, prisoners who refused to talk were accused of not having a proper attitude. Their wrists, forearms, and elbows were tied together behind their backs in an excruciating maneuver and they were hoisted into the air by another rope tied around their wrists and thrown over a ceiling beam or through a hook; they were usually left in this position for hours—to consider their attitudes. Most prisoners were kept in solitary confinement in cramped, bare cells with no lights, toilet facilities, heating or blankets. They were confined by painful leg stocks at night.

The prisoners' diet consisted of rice gruel or pumpkin soup, twice a day. This diet caused serious weight loss which led to weakness and low morale. Communication with other prisoners was forbidden and heavily punished when detected. Nonetheless, the prisoners did manage to communicate by tapping out Morse Code messages on the cell walls. Later, a "tap code," based on a five-by-five matrix of the alphabet (minus the letter "K")—first developed during the American Civil War—was used, as it was easier to remember and send. New prisoners were quickly (and secretly) taught the code and how to "get on the walls" to communicate with their fellow inmates. This communications saved the life and sanity of many despondent prisoners.

In addition to brutalizing the captive Americans, the government of North Vietnam steadfastly refused to acknowledge how many prisoners it held or who the prisoners were. Several public "shows" were staged for carefully chosen foreign press representatives or American individuals and groups opposed to the war. At these shows, apparently well-fed, well-clothed, and well-treated prisoners were brought in to meet the foreign visitors.

Some of these shows backfired on the North Vietnamese. For

example, Captain Jeremiah A. Denton, Jr., a U.S. Navy pilot, made a statement that he knew would be filmed; as he spoke, he blinked his eyes in Morse Code to spell out the word "TORTURE" over and over. (This was later detected by a sharp-eyed Navy intelligence analyst.) Lieutenant Commander Richard A. Stratton, another Navy pilot, marched about the room stiffly, staring straight ahead all the time and bowing very deeply—all to convey the message that he had been forced to be present at the "show." Many of the prisoners were more blunt (if sometimes secretive) when they could be, following the lead of Seaman Douglas B. Hegdahl. Whenever he was introduced to American war opponents, Hegdahl gave them the finger. Additionally, if he was put anywhere near food in the rooms, Hegdahl began to stuff his mouth with as much as he could get in before his captors forcibly removed him.

Later, the prisoners paid for any of their misbehavior; but they were willing to take the punishment because they knew they had embarrassed the North Vietnamese and got the word out that they were not cooperating with them, despite whatever might be said by the Communist government.

The presence of American POWs was a major war issue and continued to be so even after the peace talks began in Paris during the Nixon Administration. America adamantly demanded information about and access to its personnel held in prisons. The government of North Vietnam just as stubbornly refused to cooperate on the subject.

Based on various sources of intelligence (human sources as well as overhead photography from drones, planes, and satellites) and on information given by nine prisoners who were released between February 1968 and August 1969, many of the prison camp locations were known. Hegdahl, whom his captors thought was retarded, was ordered by the Senior Ranking Officer of his camp to accept parole if it were offered to him. This order contradicted the Code of Conduct for Prisoners of War but was done deliberately because the Americans knew something the North Vietnamese didn't: Hegdahl had a photographic memory and had remembered personal data, shoot-down or capture dates, physical condition, and locations on over 250 American POWs.

Eventually, Hegdahl was released in the last group of three in August 1969. The information he brought out with him was a veritable gold mine of intelligence.

Camp locations, including the names given by the POWs— names such as the Briarpatch, Rockpile, Skidrow, Alcatraz, the Zoo, the Hanoi Hilton (a reference to Hoa Lo) and others—were precisely plotted and photographed from overhead platforms. Data similar to that kept in Intelligence Target Folders was collected, collated and analyzed on each camp. Because of the way these prisoners had been released, the U.S. government was reluctant to say publicly what it had learned from them about the horrible conditions of those who remained behind. The fear was that such public statements would lead to harsher treatment and a suspension of further prisoner releases. Information about American POWs, however, became a very high priority for collection.

In early 1970, intelligence analysts of the U.S. Air Force's 1127th Field Activities Group, stationed at Fort Belvoir, Virginia, began developing information about two camps west of Hanoi. One was located about 30 miles west, in a place called Ap Lo. The other was not as far west, only about 23 miles; it was located inside a walled area near the provincial capital of Son Tay. In late April, reconnaissance photographs showed increased outdoor activity at Son Tay. Imagery analysts concluded that the compound was being expanded.

Then, on May 9, a technical sergeant discovered something very unusual on a photograph of the Son Tay camp: he thought he read a message from the pattern of clothes and other objects that spelled out a message based on the "tap code." The message seemed to suggest that six of the prisoners were planning to escape and wanted to be picked up in the foothills near a village eight miles southwest of the camp. The sergeant had, prior to this Saturday morning, assembled fairly conclusive evidence of the presence of 55 POWs at Son Tay. Now it appeared that several of them were planning to go over the wall and were asking for help.

During the next two weeks, briefings and meetings were held in various Air Force organizations to discuss the sergeant's analysis of the Son Tay imagery. Not everyone agreed with his analysis, although the supervisory staff of the 1127th did and they continued to push for a mission to help rescue the escaping POWs. Following a briefing to the Air Force DCSP&O (Deputy Chief of Staff Plans and Operations), a meeting was arranged for May 25 so that the analysts of the 1127th could brief an Army brigadier general named Donald D. Blackburn, who had the interesting, but obscure, title of Special Assistant to the Chairman (of the Joint Chiefs of Staff) for Counterinsurgency and Special Activities. Most people referred to Blackburn as the SACSA.

General Blackburn came to the position with many years of experience in unconventional operations. In 1942, as a young Army officer, he'd been stationed in the Philippines. When the Japanese overran Luzon, Blackburn was one of several officers who disobeyed the order to surrender. Instead, he took to the hills in the north of Luzon and established a guerrilla army among the Igorot headhunters. In the late 1950s he transferred into the Army's Special Forces; he organized and sent to Laos the "White Star" teams that trained the nascent Laotian Army. In 1965 he was selected to head the Studies and Observation Group (a cover name for the Special Operations Group, or SOG), which directed much of the unconventional war in Vietnam, particularly controlling an all-services force that conducted many of its operations out-of-country. Blackburn was later an assistant division commander of the 82nd Airborne Division. In Army circles he was a vocal supporter of unconventional operations.

Blackburn and his assistant, an Army colonel named Edward E. Mayer, Chief of SACSA's Special Operations Division, listened intently to the briefing. By now the 1127th had included in its briefing some details for how a rescue mission could be executed. These details included putting an Army Special Forces team within striking distance of the camp and sending a human agent into the Son Tay area to assess the situation. SACSA had control over assets that could, if approved, arrange for all aspects of the proposed plan. Blackburn liked the idea, for several reasons. He had other plans for special operations in North Vietnam and a success like this could boost his ability to get these other plans approved. The most important reason he liked this plan, however, was that if it could be pulled off, the rescued prisoners could speak out about conditions in the camps, unhindered by any concern for violating the "conditions of their release."

As the meeting progressed, Blackburn suggested that if they could get six prisoners out, maybe they could hit Son Tay and Ap Lo at the same time and get all the prisoners at each camp out. What a coup that would be! However, Blackburn didn't have the authority to approve such a mission; that would have to go to the Chairman of the Joint

Chiefs of Staff. As the meeting ended, Blackburn promised to get back to the 1127th after describing this brief to his boss. Soon after the Air Force intelligence analysts left, Blackburn and Mayer briefed General Earle Wheeler, the chairman. Wheeler's immediate reaction was, "Jesus Christ, Don. How many battalions is this going to take?" Nevertheless, he gave Blackburn a go-ahead to put a plan together and be prepared to brief a meeting of all the chiefs on June 5.

Over the course of the next six weeks, Blackburn and Mayer were busy briefing the Joint Chiefs, receiving initial approval and then forming a feasibility study group to look at supporting intelligence and possible operational options. The raid was given the code name Polar Circle. One of Blackburn and Mayer's most important tasks during this period was to find the correct people to tell about the operation who could support them, yet still keep the list of those who did know down to the absolute minimum. Blackburn was convinced that he would only get one chance at an operation like this and any breach of security would be disastrous.

A decision made early in the planning phase was to fly the raiding force to Son Tay in helicopters from somewhere in Thailand, probably U Dorn Royal Thai Air Base. Such a flight would require the Air Group to "thread the needle" of all the intervening anti-aircraft radar sites in order to to get in undetected. The National Security Agency provided detailed information about North Vietnamese radar networks, with blind spots identified. Precise routes would have to be plotted, the Air Group would have to fly as close to the ground as possible and an inflight refueling would be required. It was also about this time that an idea was raised to conduct a diversionary air raid over or near Haiphong, on North Vietnam's east coast, to draw all attention to it and away from Son Tay.

On July 10, Blackburn and a member of his feasibility study group briefed the new chairman, Admiral Thomas Moorer, and the other chiefs, seeking approval to begin assembling a force to carry out the raid. At this, Moorer's first meeting as chairman, several new issues were raised. Analysts were certain that the camp at Ap Lo was empty; they had no idea where the prisoners had been sent but they knew conclusively that they were gone. At Son Tay, 61 prisoners had been positively identified by name as being in the compound. Additionally, a compound south of Son Tay that looked very similar to the prison camp was identified as a "secondary school." The Joint Chiefs gave Blackburn approval to recruit a raiding force and begin training.

That following Monday, July 13, Blackburn and Mayer went to Fort Bragg, North Carolina. They had lunch with Colonel Arthur D. Simons, the Assistant Chief of Staff G-4 (Logistics), XVIII Airborne Corps, and asked him if he was interested in leading a "very sensitive mission," without giving him any further details. Simons' answer was, "Hell, yes, let's go. I don't need to know any more about it."

Blackburn was not surprised at the answer from the Special Forces legend whose nickname was "Bull." Simons' unconventional career also stretched back to World War II, when he was an original member of the 6th Ranger Battalion in the Pacific Southwest, commanding B Company. His operations with the Rangers formed the start of his legend, as he led several operations behind Japanese lines to conduct reconnaissance or raids at the direction of Sixth Army headquarters. He transferred to Special Forces in the late 1950s and eventually was chosen by Blackburn to lead the "White Star" teams in Laos. He later commanded Special Forces units in South America and Vietnam. Most of the men who served with him were willing to go anywhere he led them.

Blackburn, Mayer and Simons then discussed who else should be on the operation. At least two other names were added to the team: Lieutenant Colonel Elliott P. Sydnor and Captain Richard J. Meadows, both currently assigned to the Ranger Department, U.S. Army Infantry School, Fort Benning, Georgia. Meadows had served with Blackburn and Simons before in Special Forces assignments. He was a sergeant on a SOG team when Blackburn was chief of SOG. He led a SOG operation into Laos and captured a North Vietnamese artillery piece. For this incredible mission he received a battlefield commission, the first awarded during the Vietnam War. He later served with Simons in Panama and impressed him with his unconventional skills and his nerve. Sydnor, too, had an extensive background in Special Forces and had served with Simons before. Simons considered him one of the best Special Forces soldiers he knew.

Before leaving Fort Bragg, Blackburn and Mayer determined that training areas for their raiding force were not available there and decided to use Eglin Air Force Base, Florida, the home of the Air Force Special Operations Force. Later that same day, Blackburn and Mayer received approval from Admiral Moorer to use Eglin. A message was sent to the Air Force at Eglin to name a mission commander for what was then called "Operation Ivory Coast." The ad hoc organization that would conduct this operation was given the innocuous name "Joint Contingency Task Group" (JCTG). Before long, Air Force Brigadier General Leroy J. Manor was named to command the JCTG.

Manor had over 350 combat air missions, flying in World War II and Vietnam. He was an interesting compliment to Simons. He was quiet, yet had the same degree of efficiency as the Special Forces colonel. His recent experience in Vietnam had brought him into contact with many of the best fliers in the Air Force, both fixed-wing and helicopter. He had trained many of them to support SOG's unconventional operations.

Unknown to all the planners, on July 14 all American prisoners were removed from the camp at Son Tay. The camp's well had run dry and recent flooding from the nearby Song Con River threatened the compound with water that came to within two feet of the prison camp's wall.

In Washington on the 14th, Blackburn told Manor and Simons everything he knew about the planned raid on Son Tay. Manor would be the overall JCTG commander and Simons would be his deputy. Additionally, Simons would train and lead the raiders (Army and Air Force), while Blackburn and Mayer would run interference in Washington. That same day Manor was given a "To whom it may concern" letter from the Chief of Staff of the Air Force, ordering all Air Force commanders to give Manor whatever he needed, "no questions asked." Field training for the Army raiders was to begin on September 9. The first window of opportunity for the raid, based on weather and light data, was between October 20 and 25, since North Vietnam would be in its monsoon or rainy season until just before then. Finally, Manor and Simons were told to have their operational plan ready by the end of August.

Later that summer, after bringing Sydnor and Meadows on board, Simons addressed a theater crowded with Special Forces troopers at Fort Bragg. He couldn't tell them much except that the mission would be hazardous and there would be no extra pay for temporary duty; more than half of the initial group left. In addition to those who came to the theater, sergeant-majors of several Special Forces units on post forwarded the names of combat veterans in their units who might be good candidates for a hazardous mission.   During the next three days Simons and the raiding team doctor (an Army Special Forces-qualified lieutenant colonel named Joseph Cataldo) and several others screened and interviewed the remaining volunteers. Medical and personnel records were reviewed. The list of questions in the interview would tell those being interviewed nothing about the operation. The questions covered a wide variety of topics that were more designed to give Simons and Cataldo a chance to listen to the volunteers, assess their potential physical abilities (such as being able to carry a prisoner from the compound to a helicopter) and observe them answer questions that embarrassed them and even put them under stress. Eventually they got their final list down to 15 officers and 82 enlisted men.

Most of these men were combat veterans, although six of them were not. Typical of the kind of soldier selected was Master Sergeant Galen C. Kittleson. During World War II, Kittleson had, as a nineteen-year-old PFC, served on the Nellist Team with the Alamo Scouts, the special reconnaissance unit created by Lieutenant General Walter Krueger, commander of Sixth Army. Kittleson was on the two operations of the Alamo Scouts that liberated POW camps—the first in October 1944 at Oransbari, New Guinea, and the second in January 1945 at Cabanatuan on Luzon. He left the Army after the war but came back in during the mid-1950s and later joined Special Forces. In March 1967, while in Vietnam, he was on an unsuccessful raid to free James N. ("Nick") Rowe, a U.S. Special Forces officer held prisoner in the Delta region of South Vietnam. In 1978, when he retired from the Army as a sergeant major, Kittleson was the last of the Alamo Scouts on active duty.

Then Simons sent representatives to Eglin Air Force Base to pick a training area. As coincidence would have it, the area that was finally used was near Auxiliary Field 3 (known locally as "Aux 3" and "Duke Field"), the same area that was used to train the pilots for the Doolittle raid in 1942. Helicopter pilots and crews were picked next, from the cream of Air Force Special Operations. Manor led this effort because of his position and because he knew who most of the good Special Operations pilots and crews were. Like Simons, he knew who he wanted to take with him on an operation like this. Lieutenant Colonel

Warner A. Britton was the senior helicopter pilot and his main job was to train the air crews so they would be ready when joint training began with the Special Forces troopers, in late September.

While the training area at Eglin was being established, Blackburn convened a planning group in Washington for a solid week of making decisions and reviewing current intelligence on the target area. During this period the Defense Intelligence Agency (DIA) briefed the group that Son Tay had not seemed very active since early June. The operational planners devised long flight missions to train the helicopter crews, while the ground force was to concentrate on physical training, team and patrol drills, and marksmanship—a lot of marksmanship. Two elements, to perform logistics and armorer duties for the entire force, were selected and given broad mission orders to begin drawing equipment—or buying it or stealing it, if necessary.

Both groups, Air Force and Army Special Forces, brought a broad background to the force, but Simons intended to focus their effort. The training and drills he put them through were all based on what they would do in the actual operation, plus some contingency training, such as survival, escape and evasion, and a healthy dose of emergency medical treatment. Only four people in the ground force so far knew what the real target was: Simons, Sydnor, Meadows and Cataldo.

A cover story was developed to tell the men that would give them sufficient information about the operation but not be too close to the truth. They thought they would be raiding a village to rescue diplomats and other embassy personnel held hostage. The operational name, Ivory Coast, led them to think they were headed for Africa, and Simons let them think whatever they wanted. As time passed, many of the men figured out that they would be hitting a POW camp, probably somewhere in Southeast Asia, although none imagined it would be in North Vietnam.

Soon they began training with a collapsible mock-up of the village, made of wood and target cloth. It could be, and was, easily assembled in the dark (when satellites couldn't photograph what they were doing), and be disassembled and stored away before first light. They also met "Barbara," a detailed terrain model mock-up of the prison compound and the surrounding area with special features to simulate various degrees of available light. It had been built for them by the Central Intelligence Agency.

As the training progressed, Simons examined the raiders' actions more closely. He began to look for ways to take out the tower guards more efficiently and get part of his force into the compound quicker. The first part of the plan called for a helicopter to fire at the guard towers with M-60 machine guns, taking out the guards. The other part was difficult because of the space inside the compound; Simons thought that landing a helicopter inside would give the prisoners less time to be exposed to fire by their captors before his raiders could get to them and would also give his force the added advantages of surprise and shock.

At first an Army UH-1—Huey or "slick"—helicopter was used with a select Army crew. This aircraft was dropped because it did not have the carrying capacity needed, could not be refueled in flight and did not have the range to fly from Thailand to Son Tay. The solution to these problems was to use the HH-53 (Super Jolly Green Giant) for transport since it fulfilled all the criteria the Huey did not. The air crews then suggested crashing an HH-3 (Jolly Green Giant) helicopter instead of the Huey. Although smaller than the HH-53, it was just as rugged and durable, and its blade length would give them just enough clearance. This also meant that the HH-3 that was to be crashed would have to be destroyed, then left behind before the raiders pulled out.

Simons eventually divided his ground force into four teams, headed by himself, Meadows, Sydnor and Lieutenant Colonel Bill L. Robinson. Meadows' team (code-named "Blueboy"), the smallest, had 14 men. Their job was to be on board the helicopter that would crash inside the compound. Simons' team (code-named "Greenleaf"), of 20 men, would be next, securing a perimeter for Sydnor's team. This team (code-named "Redwine"), of 22, was the largest to land and would breach the prison outer wall with a satchel charge. Robinson led the largest team overall, 36 men. They acted as a back-up pool, and had to know what virtually every man on all the other teams did, in case any of them could not make the final operation. Some shifting did occur, although not much. For example, as late as just a few days before the raid, one of the men on one of the first three teams had to be replaced. Robinson's team also supported the other three in training in various ways, such as setting up and tearing down the simulated village and acting as the enemy force.

Obtaining the equipment and ammunition the air and ground groups needed for their extensive training and for the eventual mission was a monumental task. There were items that were in scarce supply in the military system; some had to be purchased commercially because that was the only source. Reliable ammunition was a persistent problem and not just because of the variety of weapons that would be used. So many exceptional purchases or requisitions were made that Simons was concerned that his logistics "tail" would compromise the mission.

Once most of the logistics problems had been resolved, live fire rehearsals began, both day and night iterations. On September 28, the air and ground groups began joint training. The tempo of three assaults in both day and night hours continued. When not rehearsing, the teams walked through their actions in and around the compound so that each man knew exactly where every other man was and would be. "Barbara" was a constant training aid in these sessions. By now the Special Forces raiders had fired so many rounds in so many rehearsals (between 150 and 170) that Simons was certain they knew where each round would land.

On about October 6, Manor and Simons decided their men were ready for the first real-time "full-scale, live-fire, night dress rehearsal." The air portion of this rehearsal may have been the most difficult, as would be the actual air execution on the raid, because the flight involved many out-of-the-ordinary hazards, just in reaching the target. The accompanying and re-fueling C-130s had to fly at 105 miles per hour, just over stall speed, and with 70 per-cent flaps so that the HH-3, the slowest of the helicopters, which had a top speed of about 105 miles per hour, could keep up. The A-1s and other planes that would also be flying this leg couldn't fly that slowly and so the pilots had to execute S-turns and circling maneuvers to stay with the remaining aircraft. This would have to be done at low level, all the while evading radar and in complete radio silence.

In Washington, Blackburn continued the round of briefings to receive approval for the final operational launch. Political meetings abroad by President Richard M. Nixon and his National Security Advisor, Henry A. Kissinger, continued to center around obtaining more information about or getting the release of American prisoners. While he was in Europe on a trip, President Nixon was first briefed on the Son Tay raid and he authorized the planning to proceed.

In early October the raid was postponed for a month because of political considerations. This extra time did little to improve intelligence about Son Tay. Many satellite and SR-71 photographs were still useless because of heavy cloud cover. Drones were sent sparingly for fear that too many would signal a particular interest in the area; the "take" from those missions that did go was not good. DIA analysts reported that there appeared to be less activity at Son Tay than they had seen in the past and they did not observe recent outdoor activity by the POWs. They concluded that perhaps the prisoners were being punished.

On November 1, Blackburn, Manor and Simons left for the Pacific area on a final round of pre-raid coordination. This included visits to Hawaii and Saigon, capital of South Vietnam, and a separate trip to brief Vice Admiral Frederic A. Bardshar to arrange for the Navy carriers on Yankee Station in the Tonkin Gulf to launch the diversionary raid on Haiphong Harbor. While they were gone, analysis of new photographs indicated an increase of activity at both Son Tay and the "secondary school" to the south.

On November 12, the White House gave a conditional authorization for the raid to go on the evening of November 21; a final authorization briefing for the President was scheduled for November 18. Also on the 12th, the first C-130 Combat Talon aircraft left Eglin for Southeast Asia. Two days later the raiding force departed, bound for Ta Khli Royal Thai Air Base, the designated "isolation area" for final briefings.

Blackburn and Mayer were still putting out fires in Washington. The latest involved asking the Air Force to ground all its HH-53s for a "potentially catastrophic technical problem" until further notice. This was caused by CIA requests to use aircraft earmarked for the raid. General Creighton Abrams, Military Assistance Command Vietnam commander, who knew about the raid, declined to release the helicopters to the CIA but couldn't tell them why; local CIA officers were trying to find ways to bypass Abrams and get the helicopters. On November 17, the raiding force arrived at Ta Khli.

Just before noon on November 18, Admiral Moorer briefed President Nixon on the final details of what was now called Operation Kingpin. The President approved the raid and a message to "execute" was sent to Manor in Thailand. At about the same time, Typhoon Patsy slammed into the Phillipines with winds of 105 miles per hour, and turned west toward North Vietnam.

At 4:30 the next afternoon (4:30 in the morning on November20 in Thailand) the Director of DIA told Blackburn that there might not be any prisoners in Son Tay, that the compound looked empty. A meeting was quickly arranged with Admiral Moorer. A decision was made to meet at breakfast early the next morning with the Secretary of Defense, Melvin Laird, and brief him. During the night, at just before four in the morning, Washington time, Manor notified the Pentagon that he had moved the raid up by 24 hours because of Typhoon Patsy, which was now predicted to hit North Vietnam on the night of the 21st. The raid would launch at 10:32 A.M., Washington time, unless he received orders to the contrary.

What Manor did not say in his message was that he had increased his air group. He decided to add five F-105s to escort the raiding force into North Vietnam to act as decoys once they were over the target— he hoped that they would draw the attention of anti-aircraft batteries and SAM crews. After he sent this message, Manor departed Thailand for Monkey Mountain, south of Da Nang, in South Vietnam, a communications center where he could receive traffic from the raiders at Son Tay as well as the National Military Command Center at the Pentagon; he would remain at Monkey Mountain until the raiders left Son Tay.

Blackburn had spent much of the night reviewing recent intelligence reports with DIA analysts. Human sources not specifically targeted to collect data on POWs provided information indicating that there were no prisoners at Son Tay. Infrared imagery of the compound, however, indicated activity and photographs showed that the compound had recently been enlarged. Information provided by a special source showed that information would soon be released that at least eleven POWs had died since the beginning of the year, one as recently as early November; this gave added urgency to the mission. Blackburn finally told the analysts to make one comprehensive review of what they had and to give him an unequivocal answer in the morning. At a meeting before breakfast, the intelligence analysts told Moorer and Blackburn that they had conflicting information but recommended that the raid go. This was passed to Defense Secretary Laird at breakfast and he agreed. The raid was now a "go" for the final time.

In Thailand, the raiders—who still did not know the real mission—had been going through last-minute briefings, re-zeroing weapons and holding shooting practice for two days. On the afternoon of November 20, with the entire force assembled in a theater, Simons spoke briefly to the men. "We are going to rescue 70 American prisoners of war, maybe more, from a camp called Son Tay. This is something American prisoners have a right to expect from their fellow soldiers. The target is 23 miles west of Hanoi." Simons recalled later, "You could hear a pin drop. I want to tell you it got pretty quiet. Very quiet." Then, as one, the men stood and applauded. "You are to let nothing, nothing interfere with the operation," Simons went on. "Our mission is to rescue prisoners, not take prisoners."

Simons then offered to let any man who did not want to go on the mission leave and nothing would ever be said about it. No one left. After a few more comments, Simons turned the briefing over to Sydnor. As he left the room Simons heard one of the men say, "Jesus, I'd hate to have this thing come off and find out tomorrow I hadn't been there."

Soon after, the force loaded onto a C-130 and flew to U Dorn. On Yankee Station, as the raiders flew to their jumping-off point, Navy pilots from three carriers, U.S.S. *Oriskany* U.S.S. *Ranger* and U.S.S. *Hancock* were getting the word that they would be flying missions armed only with flares. Bardshar's orders were very specific: "No air-to-ground ordnance is authorized with the exception of flares carried by Strike aircraft and the Rockeyes [cluster ammunition] and guns carried aboard the Rescue Combat Air Patrol." At the last minute, Bardshar did authorize a few aircraft to be armed with "radar-homing missiles to suppress . . . air defense batteries."

Just before 11:00 in the evening, the helicopters with Simons and his men on board took off from U Dorn. They were led by two C-130 Combat Talons. Simons told his men to wake him when they were 20 minutes out from Son Tay, then went to sleep. At about the same time a refueling C-130 left Ta Khli, and the A-1s, F-4s and F-105s left from Nakhon Phanon—all headed to Laos, flying without navigation and cabin lights and in radio silence.

Thick clouds covered the route through Laos, making visual contact difficult; at each break in the clouds the air convoy resumed formation. Over central Laos refueling began. One of the helicopter pilots later described this operation: "Well, daytime is pretty easy. Nighttime with the lights on is not too bad, but nighttime with no lights and no radio will damn near give you a heart attack." Refueling was accomplished without a hitch, the last helicopter beginning its refueling at H minus 55 minutes. At the same time, on Yankee Station, Navy jets took off to begin the diversionary raid.

As the air convoy crossed into North Vietnam, the various aircraft continued their delicate airborne dance, twisting and turning through the blind spots in the air defense radars as they approached Son Tay from the west. The final checkpoints in the flight were approaching: the Black River ten miles from the compound, Finger Lake (the raid holding area, where aircraft would go after dropping their raiders, waiting to be called back to pick up them with the freed POWs) seven miles out, and the Song Con River where it turned north, just two miles out. The air convoy began to split apart, each aircraft heading off to perform its mission.

Just as they crossed the last checkpoint, at exactly 2:17, the sky to the east over Haiphong lit up. The U.S. Navy had arrived, although the raiders could not see that far. Right on cue, one minute later the Combat Talons near Son Tay began dropping flares, lighting up that area. The A-1s dropped combat simulators, to make any forces in the area think that a fire fight on the ground was in progress. The F-4s flew screening missions for all the other aircraft while the F-105s hoped to attract the attention of SAM batteries away from the helicopters to themselves.

Major Herbert Kalen bought the HH-3 (carrying "Blueboy") toward the Son Tay compound, preparing to crash inside. Kalen maneuvered the helicopter so that the gunners manning M-60s with all-tracer rounds could take out the guard towers on the wall but ignore the tower over the gate. This tower was on top of a metal shed. They were told in training that prisoners were put in this shed as punishment, and Kalen's gunners (Kittleson on the port-side gun and another raider on the starboard gun) were to take no chances that a prisoner was in the shed.

As he cleared the trees Kalen hit a clothesline. His careful maneuvering was gone and it was all he could do to make sure he landed upright. Just before touchdown the blades cut into a 10-inch-thick tree, twisting the helicopter sharply to the right before it slammed into the ground. The mattresses on the deck inside the helicopter, to absorb the blow of the hard landing for Meadows and his team, didn't help much. A fire extinguisher was knocked loose, hitting the flight engineer (Technical Sergeant Leroy M. Wright) so hard that it broke his ankle. Meadows was to have been the first out of the helicopter, but the force of the crash landing threw Lieutenant George W. Petrie out first. Meadows followed, leading his men. When he was clear of the wreck, he knelt, holding a bull horn, as his men headed for the cells and the front gate.

The next helicopter out from the compound, piloted by Lieutenant Colonel John Allison, with Sydnor's team ("Redwine") on board, had watched the aircraft in front of him as it swung into action at 2:18. Kalen's helicopter had started down, then come back up and pulled ahead. Allison realized that Kalen had made a mistake that they had been warned about in training—he had almost landed at the "secondary school," which looked a lot like Son Tay. There was even a nearby canal that could be mistaken for the Song Con River. Then Allison noticed that Britton's helicopter (with Simons' team, "Greenleaf," on board) veered off to the right. Britton landed at the "secondary school." Allison realized that they had drifted southward. He corrected his line of flight and headed for Son Tay.

When he arrived, Allison prepared to land just as Sydnor passed the word to his men to execute Plan Green. This was one of several contingencies that Simons had planned, each presupposing that one of the helicopters with Special Forces on board did not make it. Allison told his gunners to take out the towers that should have been Britton's target and touched down. Sydnor and his men spread out quickly. Master Sergeant Joseph W. Lupyak led the way to several buildings outside the compound that housed guards—those off-shift who were sleeping, and those onshift, who were not on post but were still awake. Master Sergeant Herman Spencer moved to the compound wall and set a satchel charge. It was to blow a hole for them to enter the compound. Allison lifted his helicopter off just before the charge blew.

It did not take Simons long to realize they were in the wrong place. He could see barbed wire on the compound walls and did not expect to; he could not hear Meadows talking on his bull horn, as he had expected. As a fire fight broke out around him, Simons told his radio operator, Staff Sergeant Walter L. Miller, to recall the helicopter and tell Sydnor to go to Plan Green. Then he told his signal specialist, Staff Sergeant David S. Nickerson, to turn on a strobe light to mark the landing zone for Britton. Finally, he turned his attention to the fire fight around him.

Meanwhile, inside Son Tay, Meadows was talking into his bull horn, saying "We're Americans. Keep your heads down. We're Americans. This is a rescue. We're here to get you out. Keep your heads down. Get on the floor. We'll be in your cells in a minute."

The raiders on Meadows' team were running through the buildings, shooting at waist and shoulder level, figuring that the Vietnamese would not understand, as they hoped the POWs would, the instructions over the bull horn. Just as Meadows' radio operator called to tell Simons that they were inside the compound, a large explosion knocked Meadows down. Sydnor's team came pouring though the hole they had just blown. Sydnor quickly told Meadows about Plan Green and the two team leaders immediately began to coordinate the actions of their men.

Simons' attention was caught by a moving light in the sky and a loud *wooooosh*. At first he thought a plane had been hit but, when the light kept moving, he decided it was a SAM, streaking skyward in search of a plane. Almost as suddenly there was a big explosion at the compound. Simons knew his men did not have explosives and assumed that a stray round had probably hit some gasoline barrels. He and his men rushed into the compound (Simons was never sure how the hole in the wall they rushed through got there) and shot at everything that moved. For five minutes the fighting went on.

Back at Son Tay, Master Sergeant Thomas J. Kemmer and Petrie searched what they believed was an administration building. They ran from room to room. On the other side of the compound, Captain Udo Walther led a team through the cells, shooting several Vietnamese soldiers as they went. Outside the compound Cataldo had established a medical screening station. In the center of the compound, Meadows was hearing the same message from his various search teams: "Negative items. Negative items." It was a message he had not expected to hear; it meant there were no prisoners.

Simons was also getting a surprise he definitely did not expect:

The people his men were shooting were tall, many of them close to six feet, some taller. Some were Oriental but not all of them. They wore uniforms different from what he knew North Vietnamese soldiers wore. The fight was too hot and heavy for any of his men to determine with any success who they were fighting, but afterward they speculated that they may have been Russian or Chinese.

Simons got a radio message that his "ride" was inbound and he pulled his men back to the landing zone, fighting all the way. Britton landed under fire and stayed until he was sure everyone was on board. In the short trip to Son Tay, Simons radioed in an air strike on a road bridge between the "secondary school" and Son Tay ("insurance" he called it later), and called Sydnor and Meadows to tell them he was on the way and to revert to the basic plan. In less than a minute from pickup, he and his men landed outside the compound at Son Tay. Britton had just completed three combat sorties in less than nine minutes! His navigation error was probably one of the most fortuitous "accidents" anyone on the raid had ever seen.

As Simons and Meadows linked up in the compound there were still several small fire fights going on. When Meadows told his colonel what he was hearing from his search teams, Simons was stunned. Meadows left to conduct a thorough search of his own. When he got back, the result was the same: "Negative items."

Without hesitation, Simons told his men to prepare to move to the landing zone for extraction. He set up security and sent one of his photographers back into the camp to take pictures of the empty cells. Then he told Miller to radio for an air strike on a bridge north of the compound and to bring two of the helicopters back from the holding area. Nickerson fired a flare to aid the pilots in finding the correct compound. Fourteen minutes had elapsed since the gunners hosed the guard towers.

When Britton landed outside the compound, Simons sent Meadows' team (without Meadows), Kalen and his crew, and part of Sydnor's team (26 in all) on board. Britton lifted off and Simons radioed for the next helicopter. As the remaining 33 raiders waited for Allison to bring in his helicopter, Meadows went back to the compound one last time to affix demolition charges to the crashed HH-3. Then he dashed back out to wait for Allison.

In the sky overhead, SAMs continued to fly, but they hit nothing.

One came in the general direction of Britton's helicopter but his crew chief spotted it and yelled a warning into his radio. Britton banked sharply to avoid the missile. While they waited, the raiders on the ground spotted several trucks headed in their direction. They fired several LAW (Light Anti-tank Weapon) rockets at the trucks, taking out this last threat to the raiders. When the helicopter landed, the raiding force Marshalling Area Control Officer counted everyone aboard (as he had for the first helicopter) and passed the count to Simons. Allison pulled his helicopter off the ground twenty-seven minutes into the raid. Six minutes later the HH-3 blew up inside the prison camp.

Everyone (Army Special Forces, Air Force crews, and Navy pilots) made it back alive. Two raiders (Kalen's flight engineer and one Special Forces raider) were wounded and one F-105 was shot down after the raid, but one of the HH-53s pulled the crew out and brought them back. By the time they all arrived in Thailand, General Manor had returned from Monkey Mountain and he greeted them as they landed.

Several years later, details of a program known as Operation Popeye (and several other names) were revealed. Popeye involved attempts to alter weather patterns over North Vietnam. Specifically, it tried to increase the seasonal rainfall in northeastern Laos, hoping to cause heavy flooding in western North Vietnam. Most of the Popeye missions in 1970 occurred between March and November.

The Son Tay raid's After Action Report (written by Manor, who did not know about Popeye) said that five years' worth of typhoons hit the area of North and South Vietnam and Laos in the two months prior to the raid. From Blackburn down, no one involved in the raid planning was aware of Popeye. No one really knows for certain what influence Popeye had on the flooding that contributed to the decision by the North Vietnamese to move the prisoners out of Son Tay in July 1970.

It had been a near-perfect raid. Even the landing of Simons and his team in the wrong place was somehow right. The raiders had achieved complete surprise and, in Simons' words, the "attack was pressed with great violence—because surprise doesn't work if you don't use violence and speed." And yet, they came away emptyhanded.

Had they failed? There is a school of thought that contends they did not fail, pointing to the precision of the raid itself and, more important, from their perspective, what happened to the POWs who remained in North Vietnam. Conversely, others believe the raid did fail, pointing to the failure of intelligence to give complete answers for the planners and raiders. Part of the intelligence failure, they say, was relying on merely technical means to collect information, and then limiting these means to mostly imagery. No human source was targeted to collect information about Son Tay and there was no apparent attempt to use signals intelligence except to map the radar blind spots. This school also blasts the more than six months that elapsed between the imagery with the "come and get us" message and the raid itself, while the planners tinkered with the plan to make it perfect. They even fault those who reviewed the plan, up to the President, for failing to ask tough questions and demand tough answers. It seemed they had forgotten the words of General George S. Patton, Jr., who said "A good plan violently executed now is better than a perfect plan next week."

What happened to the POWs as a result of the raid is that their captors realized their vulnerability. To compensate, they pulled all the prisoners from the outlying prison camps into one place: Hoa Lo prison, renamed Camp Unity by the POWs. Whereas, before, many of these men had spent years in solitary confinement, now they were crowded 20 or more into a room. Now they could organize and act as a group; they could help one another, especially the sick or despondent; they were a military unit once more—the Fourth Allied POW Wing. It would still be more than two years before they were freed, but the prisoners knew about the raid on Son Tay—and they were proud and glad that General Manor, Colonel Simons and their men had tried to get them out.

19

Delta Force at Desert One

BY RICHARD L. KIPER

On the morning of April 25, 1980, President James Earl Carter announced to the American people the failure of a military mission to rescue fifty-three Americans held hostage in the United States Embassy in Tehran, Iran. During that mission eight American servicemen died, one EC-130 airplane and one RH-53D helicopter were destroyed, six RH-53D helicopters were abandoned and the military prestige of the United States was shattered. What had begun as a demonstration of American resolve in the face of international banditry had ended in a blazing inferno in the Iranian desert. "Operation Eagle Claw" had failed.

The takeover of the U.S. Embassy in Tehran on November 4, 1979 by radical militants, backed by the Iranian government, presented the United States with an unprecedented dilemma. From that moment until all captives were released 444 days later, the hostage crisis commanded front-page headlines. Television broadcasters nightly enumerated the number of days the crisis had persisted. To the American public the Carter administration appeared impotent.

What the American people did not know was that two days after the seizure of the embassy, on November 6, Dr. Zbigniew Brzezinski, the president's National Security Advisor, had directed Secretary of Defense Harold Brown to develop a plan to rescue the hostages. For the next five months a select group of officers assembled in the Special Operations Division of the Joint Chiefs of Staff (JCS) to develop and, eventually, to implement "Rice Bowl," the planning phase of the rescue mission, and "Eagle Claw," the execution phase.

Such a raid had not been attempted since the unfruitful Son Tay rescue mission in November 1970 to free American prisoners of war held in North Vietnam and the May 1975 attempt to rescue thirty-nine American seamen aboard the *Mayaguez* (at a cost of forty-one U.S. servicemen killed and fifty wounded). Despite growing indications that terrorism was becoming a favored weapon of radical nations and groups, the United States had been slow to create an effective force to counter terrorist activities. Not until November 1977 did the Army formally authorize the creation of 1st Special Forces Operational Detachment Delta—the Delta Force—with the mission of combatting terrorism.

Airlift for counter-terrorist operations was a major shortcoming because the Air Force special operations capability consisted of only fourteen MC-130s scattered worldwide, several AC-130 gunships and a few new HH-53 helicopters just coming off the production line, although older H-53 helicopters were in the rescue fleet. The Army had no special operations aviation capability despite having a sizeable helicopter force.

No standing joint command and control headquarters existed to oversee multi-service special counter-terrorist operations. A contingency plan existed for deploying counter-terrorist forces overseas, but only a small specially trained force, Blue Light, existed to deploy. Within the JCS organization, the Special Operations Division (SOD), a small joint organization directed by the Army's Colonel Jerry King, existed only as a staff advisory body, not as an operational planning headquarters. The Current Operations Division of the JCS had responsibility for counter-terrorism; however, the SOD was assigned the mission to plan the rescue.

To augment King's planning staff, Army and Air Force special operations veterans were attached to the SOD. Major General James Vaught, a decorated combat veteran who also had extensive Pentagon experience, was appointed to create and then command a Joint Task Force (JTF). Colonel James Kyle, an Air Force officer with special operations experience in Vietnam, became the air component commander for the fixed-wing aircraft. Marine Colonel Chuck Pitman, a helicopter expert who also had Vietnam experience, was given the responsibility of assembling the helicopters and pilots for the mission. Colonel Charles Beckwith (another Vietnam veteran whose nickname was "Chargin' Charlie"), the Delta commander, was responsible for training the force to assault the Embassy and Foreign Ministry in order to rescue the hostages.

Eagle Claw was a raid which Army Field Manual 100-5, *Operations*, defined as "a limited objective attack into enemy territory for a specific purpose other than gaining and holding ground." JCS Publication 1-02 further defined the raid as a special operation "conducted by specially trained, equipped, and organized DoD [Department of Defense] forces against strategic or tactical targets in pursuit of national military, political, economic or psychological objectives." The raid attempt, in pursuit of a national objective, would have to be characterized by the use of clandestine techniques, would be independent of support from conventional forces at the moment of execution, and would involve a high level of risk. Operational security would be paramount. Failure could have serious domestic and international consequences.

Less than four years previous, the Israelis had performed a spectacular hostage rescue at Entebbe, Uganda, which served as an inspiration to U.S. planners due to its flawless execution. The American hostages, however, were not being held in a terminal at an international airport near an open runway; they were held in a walled compound within Iran's heavily-populated capital. Planners essentially had a three-phase operation to consider: getting the rescue force into Tehran, locating and gaining control of the hostages and extracting the hostages and the rescue force. Lack of detailed operational intelligence concerning the number of guards and the exact locations of hostages; locations and operations of air defense systems; extreme distances from friendly bases; and the presence of hostile forces were obstacles to be overcome before an operation could be attempted.

Under the supervision of Vaught and King, Kyle and the other aviators began planning a mission utilizing helicopters to insert Delta and extract the hostages. These vertical-lift aircraft provided a means to introduce the rescue force without the dangers of a parachute landing and to withdraw the hostages from a small area within the city of Tehran. Distances, however, made the use of helicopters unfeasible unless they could be refueled, probably multiple times. The only launch sites that could be guaranteed were Diego Garcia in the Indian Ocean and Egypt—a 5,000–6,000 nautical mile round trip to and from the Iranian capital. Other ostensibly friendly countries in the region were ruled out either for political or operational security reasons.

Refueling the helicopters would require MC-130 Combat Talons or HC-130 tankers, but a night-time blacked-out aerial refueling operation deep in Iranian airspace was not considered practical because of altitude requirements, pilot inexperience and technology limitations. Furthermore, the C-130 aircraft would require aerial refueling, thus enlarging the aerial force involved in the mission. Although the C-141A transport had greater range, it could not be refueled in the air, and the C-5, although air-refuelable, lacked the ability to penetrate hostile airspace at low levels. Unless the planners could find a launch site closer to Iran, or could devise a means to refuel the helicopters, the use of those helicopters would be impossible. (In April 1980, C-141B air-refuelable transports would become available, but too late to influence mission planning.)

Navy planners proposed using an aircraft carrier to launch the helicopters, and planning quickly focused on that concept. A carrier task force normally operated in the Indian Ocean. As RH-53D Sea Stallion mine-sweeping helicopters were an integral part of that task force, those helicopters could be launched without creating a significant security problem. The Sea Stallion could be stored belowdecks, could be equipped with external fuel tanks and had sufficient cargo capacity for the Delta soldiers and rescued hostages. It did not, however, have an aerial refueling capability. Providing the additional fuel to get the helicopters to Tehran and back became the critical factor. But because the carrier launch appeared to be the most practical tactic, on November 28 six RH-53D helicopters flew from Diego Garcia to the U.S.S. *Kitty Hawk* as it sailed near the island. These six helicopters were then flown to the U.S.S. *Nimitz* when it replaced the *Kitty Hawk*. *Nimitz* had picked up two other RH-53Ds en route, so eight were available for the mission.

A primary concern was pilots for the helicopters. Navy pilots were familiar with the RH-53 and carrier operations, while Marine pilots were familiar with extended overland missions and operations from unimproved airfields. As the RH-53D was a Navy helicopter, Navy Captain Jerry Hatcher was assigned the task of selecting pilots and crew chiefs. Colonel Pitman would provide Marine co-pilots and door gunners.

Because aerial refueling of the helicopters was not possible, initial plans involved air-dropping fuel bladders, pumps and hoses into Iran and establishing a refueling point. The helicopters, with Delta aboard, would launch from the carrier, fly to the refueling point, land to refuel, and then move to a hide site near Tehran. Considering the number of hours of darkness available, any option that required refueling the helicopters inside Iran necessitated a two-night operation: one to fly from the carrier, refuel and move to the hide site; one to assault the Embassy and Ministry of Foreign Affairs where the Americans were being held, rescue the hostages and exit Iran.

While planning for the air portion of the mission continued, Beckwith and Delta Force were developing a ground tactical plan to assault the Embassy and Ministry, locate and rescue the hostages (which could involve searching fourteen buildings) and get everyone out safely. Using scale models of the compounds, television news film, and information from people recently stationed in Tehran, the Delta operators developed a plan which required seventy men for the mission. The assault force would grow to 118 as the plan was refined.

From the hide site established on night one, Delta could infiltrate the city by vehicle on night two, attack the two compounds, gain control of the hostages and await extraction by the helicopters from either the Embassy grounds or a soccer stadium across from the Embassy. The helicopters would transport Delta and the former hostages to an airfield where all would be loaded on transport aircraft for movement out of Iran. With this general plan in mind, solving the refueling problem, locating a hide site and identifying an extraction airfield became top planning priorities. Beckwith estimated six helicopters would be required to carry Delta and the hostages.

To secure the refueling point on night one and the extraction site on night two, Lt. Colonel Sherman Williford was directed to develop a plan utilizing one Ranger battalion.

Although the assaults would take place late at night, Beckwith was concerned about crowd reaction to the inevitable noise coming from the Embassy. To discourage intervention, air planners proposed having two AC-130 Spectre gunships orbit the area. These aircraft, equipped with 20mm, 40mm and 105mm cannon, could lay down a circle of fire with pinpoint accuracy around the assault objective to prevent Delta from being overwhelmed from outside by screaming militants.

With a feasible skeleton plan having been developed, training of the aircrews began in earnest. As all flight operations were to be conducted without external lighting, airplane and helicopter crew members were issued night-vision goggles (NVGs), which allowed the pilots literally to see in the dark. None of the crew members had trained with them previously.

In December, Vaught conducted the first training exercise, which included six helicopters carrying Delta and an air drop of the fuel bladders near Yuma, Arizona. The helicopter pilots had difficulty keeping formation and had problems landing in the blowing sand. Two MC-130s dropped ten bladders. Parachutes failed to deploy and seven 500-gallon bladders splattered across the desert. The next night the helicopter pilots had difficulty navigating, but a change in the rigging procedures resulted in a successful parachute drop of the bladders. Because the bladders, pumps, and hoses were dispersed upon landing, however, Delta had difficulty assembling the equipment and moving it into position to refuel the helicopters.

Disillusionment both with the helicopter pilots and the air delivery scheme led the JTF leadership to look for other solutions. Later that month a possible solution to the helicopter refueling problem appeared in the form of a 3,000-gallon bladder which could fit into a C-130 cargo compartment. If a suitable landing strip could be located in the Iranian desert, the airplanes could land and the fuel could be pumped directly to the helicopters, thus eliminating the problems of the air drop. EC-130 Airborne Battlefield Command, Control and Communications aircraft were equipped with a moveable pallet which could accommodate the bladders. Three of these aircraft soon joined the force.

Intelligence personnel, utilizing aerial photography, located several promising sites for the operation. The Shah had constructed an airfield at Manzariyeh, fifty miles south of Tehran, but it was unused in 1980. It appeared to be perfect as an extraction site after the hostages were moved out of the city. Planning then centered on employing C-141 aircraft flying from Daharan to lift the hostages out of Manzariyeh.

Intelligence personnel also located an area that appeared suitable for the night-one refueling operation. The area was 220 miles from the proposed hide sites for Delta and the helicopters, appeared to be able to support the weight of the aircraft, but had a road running through the area. It was determined to be the best available location. In early March, Air Force Major John Carney, with two other personnel, flew in a Twin Otter (a small, short-runway plane) to the landing area, determined it was suitable for the fuel-laden ÃC-130s, and planted remote control lights to illuminate the landing strip for the arriving aircraft. The CIA pilots on this mission, veterans of Air America in Southeast Asia, took advantage of Carney's time on the ground in hostile Iran to take naps, while the major used a small motorcycle to inspect and prepare the strip. The location was designated Desert One.

In late March the air planners briefed Vaught on a plan to use the island of Masirah off the coast of Oman as a launch site rather than launching directly from Egypt or Diego Garcia. Impressed, Vaught approached General David Jones, Chairman of the Joint Chiefs of Staff. Use of Masirah reduced considerably the flight time for the Delta troops and eliminated all aerial refueling except for one after the MC-130s and EC-130s exited Iran on night one. In early April Vaught received approval to use Masirah.

Throughout the planning phase, air crews and ground operators continued to rehearse, although none of the rehearsals would include all the aircraft involved in the mission. Initial training focused on each unit's requirements: low-level formation flying using NVGs for the helicopter crews, blacked-out landings on dirt airstrips for the C-130 crews, airfield seizure and security training for the Rangers and compound assault procedures for Delta. In mid-December all components held another joint rehearsal using three C-130s, one AC-130, and six helicopters. The Rangers would secure the refuel site and Delta would transfer from the MC-130s to the RH-53Ds. Two helicopters aborted during the rehearsal for mechanical reasons. The rehearsal for night two indicated additional training was required to decrease time on the ground and to account for all personnel. Nine helicopter pilots were replaced after these rehearsals. Rehearsals in January, February and March showed considerable progress.

On board the *Nimitz* problems were occurring with the eight helicopters designated for the mission. The Navy was not flying the aircraft in accordance with the JTF commander's desires and there were insufficient parts available to maintain them. These shortcomings were soon corrected. Experience from the rehearsals convinced the planners that eight helicopters were required to launch from the carrier to ensure six would be available for the night-two portion of the operation. With fewer than six, the mission could not continue.

A final rehearsal in April involving two C-130s and four H-53s revealed no problems. Delta and the Rangers were not included.

To remedy the problem of lack of intelligence, in December a CIA agent arrived in Tehran and was soon joined by an Iranian agent. These men scouted proposed locations for the hide sites, reconnoitered routes from these to the city, observed the guards at the Embassy and Ministry and purchased trucks to move Delta into the city. They then rented a warehouse to hide the trucks. A few days before the mission, retired Army Major Dick Meadows (a Son Tay veteran) and three other servicemen were infiltrated into Tehran to provide Delta with last-minute intelligence and guide the assault force to the targets.

At a meeting of the National Security Council on April 11, 1980, President Carter approved the rescue mission. Aircraft began moving three days later. For several months, planners had established a pattern of moving mission-similar aircraft through Egypt to reduce the notice that might be taken of the arrival of the actual mission aircraft. On April 16 Beckwith briefed Carter on the details of the impending raid. By April 23 all aircraft, maintenance and medical personnel, Rangers, and Delta were in position at Masirah and Wadi Kena, in Egypt. The *Nimitz* was off the coast of Iran. As finalized, the plan was:

On night one, three MC-130s carrying Delta, combat controllers, and a Ranger security element, and three EC-130s carrying fuel bladders would depart Masirah for Desert One. Eight RH-53Ds would launch from *Nimitz* to link up with the C-130s in the desert. After Delta off-loaded, two of the MC-130s would depart. After refueling, with Delta aboard, the helicopters would fly the operators to a hide site and then continue to a separate site to hide the helicopters. The C-130s, with the Ranger teams, would then return to Masirah.

On night two, four MC-130s with Rangers to secure the Manzariyeh airfield, would depart Wadi Kena along with four AC-130s to provide security over the Embassy and at Manzariyeh. Two C-141s would depart Daharan to extract the hostages and the rescuers from Manzariyeh. Meadows and his agents would link-up with Delta at the hide site and transport drivers for the trucks to Tehran. The trucks would then return to pick up the other Delta troops. The Ministry assault team would travel separately from the Embassy team. Upon reaching the Embassy, the assault team would eliminate the guards outside the compound, scale the walls and locate the hostages. Another team would cover the withdrawal to the soccer stadium if the helicopters could not land on the Embassy grounds. The helicopters would move the hostages, rescue force and agents to Manzariyeh for extraction by the C-141s.

At 6:05 p.m., April 24, the lead MC-130, piloted by Colonel Bob Brenci, departed Masirah for Desert One. At 7:10 p.m. the other C-130s were launched. The helicopters had departed *Nimitz* five minutes earlier. At 8:30 p.m. and 9:00 p.m. the lead Combat Talon flew through dust clouds which obscured visibility but did not affect flight operations. An attempt to warn the helicopters was unsuccessful due to communications difficulties.

At 9:00 p.m. one helicopter landed when a blade warning light flashed. The crew abandoned the aircraft and was picked up by another helicopter which continued toward Desert One. All seven helicopters navigated through the first dust cloud, but at the second the lead aircraft landed to wait for the cloud to pass. As all aircraft were maintaining radio silence, only one other helicopter landed because the pilot had seen the lead aircraft leave the formation. The two helicopters which had landed continued the mission after the dust cloud cleared. All others continued for one hour and fifteen minutes. At that point the helicopter carrying Pitman had instrument trouble and turned back, eventually landing on the *Nimitz*. The helicopter force was then down to six, the minimum required for the mission.

At 10:00 Carney, on board Brenci's airplane, turned on the remote-controlled lights at Desert One, and fifteen minutes later the MC-130 landed. It touched down so heavily that Beckwith, in the cargo section, was thrown under a jeep. His reaction was "Great landing! That wasn't bad at all!" The Ranger team rapidly deployed to establish roadblocks. They were immediately confronted by a bus loaded with Iranian passengers. The bus was stopped with warning shots and the passengers detained. Major Tyrone Tisdale, a linguist, recited the poetry of Omar Khayyam to soothe the terrified bus riders.

Five minutes later a fuel truck drove down the road, failed to stop, and was destroyed by a Ranger anti-tank weapon. The fire ball could be seen for miles. The driver escaped in a truck that was following the tanker. (Speculation was that they were smuggling gasoline because of their movement at night.) By 12:30 a.m. the other five C-130s were on the ground at Desert One. The first two MC-130s then departed. Finally, the six remaining helicopters landed. Some of the Marine pilots, still shaken by the sandstorm they had flown through, suggested aborting the mission. Beckwith's subsequent confrontation with Lt. Colonel Ed Seiffert (now in command of the choppers because Colonel Pitman had had to turn back) was loud and almost physical.

Delayed by the dust and ninety minutes behind schedule, the six helicopters were on the ground and refueling when one shut down its engines. A hydraulic pump (one of two such systems on the HH-53) had failed and the senior helicopter pilot at the site determined the aircraft unsafe to fly, particularly because of the massive load the chopper would have to carry. Word was flashed from Desert One to Vaught in Egypt and then to the Secretary of Defense, who informed President Carter. With only five helicopters flyable, the mission had to be aborted. The frustration level among the Americans on the ground in Iran seemed unbearable. But it would get worse.

One EC-130 had completed its refuel mission and, low on fuel itself, was directed to depart as soon as the helicopters moved from behind it. The helicopter to its left rear rose, began to slip left to clear the airplane, then suddenly turned ninety degrees and sliced into the EC-130. Fuel-laden tanks burst into flame and ammunition exploded as Air Force and Marine crewmen and Delta soldiers scrambled to escape the inferno. Lieutenant Colonel Jerry Uttaro, piloting the EC-130 next to the conflagration, quickly taxied his airplane away from the fire to escape the fragments flying from exploding bullets, grenades and missiles.

Three helicopters near the explosion received extensive damage from the blast and were abandoned. The fourth had the inoperable hydraulic pump and the fifth remaining RH-53D had not completed refueling. All had to be abandoned. Delta, helicopter crewmen, and the Ranger teams scrambled to board the remaining three C-130s. A radio message was sent to Major Meadows, waiting inside Tehran, advising him of the situation. He and the other agents were left to get out of Iran as best they could. All were able to do so. At 3:00 a.m.,

April 25, the last C-130 lifted off from Desert One.

There was no question of the Americans recovering their eight dead because no one could approach the blazing EC-130. When the fire cooled, Iranians arranged the charred corpses in a line in the sand for the benefit of the world press corps. The abandoned helicopters (some left behind with secret documents revealing the mission plan) were not destroyed since the resultant explosions would have blocked the airstrip, preventing the C-130s from taking off. Or, as Colonel Kyle said after making the decision not to destroy the aircraft, "The way our luck has been running, we'd probably blow ourselves to smithereens."

In the wake of the debacle followed congressional hearings, questioning in newspapers and magazines, and the official military investigation by a commission headed by retired Navy admiral James L. Holloway. The commission concluded that the raid failed primarily because of overly stringent operational security procedures. These procedures precluded a full-scale dress rehearsal, which might have revealed some of the dilemmas encountered at Desert One—such as the confusion inherent in maneuvering helicopters and airplanes in the midst of deafening aircraft engine and blade noise and blowing sand, and the difficulty of identifying key members of the chain of command. Restrictive communications security measures and lack of adequate secure communications equipment prevented the lead MC-130 from informing the helicopter formation of the dust cloud, and also kept the two helicopters that aborted the mission en route to Desert One from alerting the remainder of the force of their status.

Clearly questionable were the decisions to abandon one helicopter because of the blade warning indicator without actually checking for a crack in the blade, Colonel Pitman's decision to return to the *Nimitz* rather than continue the mission, and the decision not to fly the helicopter with the faulty hydraulic pump by using its backup system. The loss of those three helicopters doomed the operation because, as determined beforehand by the on-site mission commander, six were required and only five were then available.

Helicopters, however, have been described as "ten thousand moving parts all working against each other," and beyond their autorotation, have no gliding capability. When they fail, they crash and no warning light can be disregarded. For the mission commander to have overruled his own pilots might have resulted in a more spectacular disaster, within the city limits of Tehran. Still, the abortion of Eagle Claw because of a shortage of crucial equipment is a fact that many participants found difficult to accept. During the storm of criticism and analysis that raged after the failed mission, a high-ranking Israeli officer had a painfully stark response for a journalist seeking his expert insight. "The United States doesn't have enough helicopters?" he asked.

But what also followed in the wake of the failure was a reassessment of America's Special Operations Forces (SOF). Since 1980 vast improvements have been made in command, control, communications, airlift, training, manpower, support functions and funding. Creation of the United States Special Operations Command provided a unified joint headquarters for all SOF. The ad hoc arrangements faced by Vaught, King, Kyle, Beckwith and the others are no longer an impediment in special operations.

Could the raid have succeeded? A raid is an inherently high-risk endeavor and this mission in particular had enormous implications for the United States. As Colonel Kyle stated, "This was our Super Bowl." The Holloway Commission concluded: "It was the ability, dedication and enthusiasm of these people who made what everyone thought was an impossibility into what should have been a success. It was risky and we knew it, but it had a good chance of success and America had the courage to try."

20

The Future of Raids

BY FRITZ HEINZEN

July 3, 1976. The quiet night that enveloped the Entebbe airport near Kampala was little disturbed by the aircraft making an approach to a runway adjacent to the new terminal. Those on guard at the airport's old terminal heard little, if any, noise. Besides, an aircraft had departed only thirty minutes earlier, so a late flight would not appear so unusual. It was almost 2300 hours and the hostages from Air France Flight 139 (who had been at the airport since June 28) had been no problem to handle from a security standpoint. They were now sleeping, unaware as to whether the 4th of July would bring the further torment of waiting for release through negotiations, or possible execution as another terrorist deadline was reached.

It had been a frantic few days for all involved, starting with the hijacking on Sunday the 27th. Flight 139, an A300 airbus, was hijacked on route from Lod Airport in Israel to Paris after its Athens stopover. The four hijackers claimed membership in the Popular Front for the Liberation of Palestine (PFLP); two were Palestinian, two were German members of the Baader-Meinhoff gang. The plane first flew to Benghazi, Libya for refueling, then it proceeded to Uganda's Entebbe Airport, arriving at 0300. Once at Entebbe the four hijackers were joined by six PFLP comrades. Later on that Monday, Idi Amin, Uganda's dictator, made his first appearance and claimed to be brokering a deal for the hostages' release.

On Tuesday the terrorists made their demands known. They ordered the release of 53 imprisoned terrorists, 40 in Israeli jails and 13 confined in four other countries. At this point the Israeli military was making sketchy contingency plans, although few seriously thought a rescue mission was likely, or even possible. Over the course of the next few days, negotiations proved problematic. Information gleaned from freed hostages and telephone conversations with Amin clearly indicated that the Ugandan strongman was collaborating with the terrorists.

By Thursday (July 1) everything had become clear. The hijacking was specifically directed against the Jews. Israeli citizens had been segregated from the other hostages on the second day of the ordeal. On the 30th of June, 47 non-Israeli passengers were freed, and their release was followed by that of 100 French hostages on Thursday. The remaining 106 hostages—Israelis, non-Israeli Jews and the flight crew—were faced with extinction as the July deadline approached.

Israel had to act. The hijacking, rather than being an "international" criminal incident, was obviously an attack on the Jewish people, part of the ongoing war that been waged against the Jewish state since its creation.

In 1972, Israeli athletes had been captured and then murdered at the Olympic Games. Tel Aviv's Lod Airport was then the site of a massacre of pilgrims—27 killed. The year 1973 witnessed the brutal Yom Kippur War, which exposed Israel's vulnerabilities as no military action had before. The tiny country—with about 1.2 percent of the population of the United States—suffered 2,838 dead and 8,800 wounded in that conflict.

After the Yom Kippur War terrorists continued to batter Israel and its citizens, both at home and abroad. On May 15, 1974 Israelis witnessed the death of 26 innocents at Ma'alot (20 of whom were school children, another 60 injured). Many of the casualties came during a botched rescue mission. In the wake of this failure and the bloody, inconclusive 1973 war with its compromise peace settlement, many Israelis questioned their military's ability to protect the nation. The litany continued through 1975 and into 1976 at the hands of terrorists armed with automatic weapons or bombs. Israelis began to fall prey to self-doubt, a potentially fatal development for a nation surrounded by hostile and far more numerous enemies.

Now Israel, as the guarantor of the safety of the Jewish people, appeared unable to do anything about nearly one hundred innocent Jewish civilians being held by a dictator over 2,200 miles away. The

German and Palestinian terrorists, as well as their patron, Amin, must have taken pride in the success of their action for those few days, while Israel seemed helpless to respond.

However, appearances are deceiving. The troops aboard the lead C-130 transport and its three companion craft were about to pull off one of the greatest raids in history, and in the process lay the groundwork for the future of military raids. The ground portion of "Operation Thunderbolt" was about to commence. (Three common names for the operation are used interchangeably here. They are the Entebbe raid, Operation Thunderbolt and Operation Jonathan, the latter a renaming of the operation after its fallen commander, Jonathan "Yoni" Netanyahu, brother of Israel's current prime minister, Benjamin.)

Eighteen short hours of intense rehearsals were now put to the test. At 2300, after seven and a half hours of flight, the first C-130 unloaded its deadly cargo. In the first plane were 35 commandos, 52 paratroopers, the infamous Mercedes, two Land Rovers, and the operation's overall commander General Dan Shomron and ground commander Yoni Netanyahu. Netanyahu in the Mercedes and the two Land Rovers drove for the old terminal. Ineffectual fire from two Ugandan guards failed to halt the convoy.

Once at the terminal, three assault elements worked with ruthless efficiency, gunning down terrorists and Ugandan guards. In less than three minutes from the time of the lead aircraft's landing, four terrorists were dead and the hostages rescued. Unfortunately, in the assault Netanyahu was shot in the chest and mortally wounded, and three hostages died in the exchange of gunfire.

Over the next few minutes additional troops and vehicles (including armored personnel carriers) offloaded from the C-130s and engaged the Ugandans, who by now were beginning to put up some resistance. By 2315 the area around the old terminal was secure and all seven terrorists at the scene were dead. At this point the fourth aircraft landed and the evacuation began. The hostages departed at 2352, less than an hour after the arrival of the lead aircraft! The last aircraft to depart did so at 0040, but only after 10 MiG aircraft of the Ugandan Air Force were left blazing. Only a British–Israeli national named Dora Bloch, who was in a Kampala hospital at the time of the raid, was left behind. Tragically, she was later killed by Amin's state police.

Forgotten over the years is the firestorm of protest in the wake of the raid. Third World and Communist diplomats and international lawyers bemoaned and condemned the Israeli action. The UN Security Council met five times between July 9 and July 14 in stormy sessions. Over the years I have written three times on the legal and moral aspects of the Entebbe raid. Occasionally I have needed to look over these writings and find that I never cease to be amazed at the hypocritical posturing displayed in the fiery condemnations. In fact, not just for its successful execution but for its nobility of purpose, Entebbe was one of the greatest raids of all time.

While the Entebbe raid was of obvious importance to Israel, it was also of marked importance to future operations and an inspiration for raids that have followed (notably the 1980 attempt by the United States to rescue its hostages in Iran). There are several points to make about future raids, some relating specifically to my consideration of Operation Thunderbolt, while others not necessarily derived from it.

Almost all the raids in this book took place in the context of a declared war or a conventional conflict. Operation Thunderbolt took place in a time of peace. There was of course "a war" between the West and terrorists, but note how loosely we must use that term. One side had no organized state structure, the other side was a disparate collection of states in which there were some organized ties, such as membership in NATO and the European Community, and many not so formal ties, e.g., a European heritage. As such, there were no declarations of war, no organized lines of battle and few operations of a standard military nature.

There was no traditional state of war between Israel and the hijackers (German and Palestinian terrorists) or between the Israelis and Idi Amin and his soldiers. In the future most raids may also take place in "peacetime" or at least in the setting of an undeclared war. While there will be raids during formal, declared wars, there will likely be far fewer such conflicts compared to the "twilight" wars so prevalent at the end of the twentieth century as obsolescent states collapse and their former citizens seek new allegiances along ethnic, religious and tribal lines that often know no formal and recognized geographic boundaries.

Entebbe was the prototypical rescue mission carried out in an atmosphere of quasi-war. Witness events in Lebanon, Somalia, Liberia, Zaire, and the former Yugoslavia and Soviet Union. Old orders in these states have collapsed, and this disintegration means the end of civil authority and the potential for the seizure of foreigners. The U.S. undertook "Operation Eastern Exit" (January 1991) to evacuate U.S. embassy personnel in Mogadishu for precisely these reasons. American citizens and diplomats, and nationals from over 30 other states, were caught up in the collapse of authority in Somalia and needed to be rescued. Fortunately the helicopter carrier *Guam*, and the dock-landing ship *Trenton*, and their embarked Marines and SEALs were available to conduct the rescue after a several-day voyage from their anchorage off Oman. CH-53Es were used to insert the forces, and these helicopters plus CH-46s later undertook the extractions. Such a mission now has its own acronym—NEOs or "non-combatant evacuation operations."

Of course, such operations for the protection of diplomats and/or other citizens will also occur against states not so disintegrated. The U.S. operation ("Eagle Claw") that ended so tragically in April 1980 at "Desert One" in Iran was intended to free U.S. diplomats held hostage by Iranians, who, while calling themselves "students," clearly had the backing of those who ruled the country. (It is worth noting that this happened shortly after the chaotic transfer of power in that country, and in an environment in which authority had not been clearly reestablished.)

Note also that raids will be necessary against adversaries operating in a third party's state. Little more than a year after Israel's raid into Uganda, on October 18, 1977 West Germany's Grenzschutzgruppe 9 (GSG 9) conducted a somewhat similar operation to free a Lufthansa Boeing 737 held in Mogadishu, Somalia.

Quite possibly a third-party raid will be conducted on one's own soil—in which neither the hostages nor the perpetrators are natives. On May 5, 1980 a British Special Air Service (SAS) detachment assaulted the Iranian embassy in London to free hostages held by a little-known terrorist group.

My second main point is that although most of the raids discussed in this book were tactical in nature, their impact was, at best, operational. (There are a few notable exceptions such as Cádiz, the

Congo and Desert One.) As noted above, Operation Thunderbolt, although a tactical engagement, had strategic implications for the future of the state of Israel. Future raids will increasingly carry such importance.

As was true at Entebbe, the stakes may well be the survival of a nation's morale, or "soul." However, the threat may be even more palpable. This is due to the nature of the danger that terrorists, rogue states and other actors on the international scene may present in the future.

The threat of weapons of mass destruction will bring increasing opportunities and pressures for raids. There are ever larger numbers of weapons available to both states and rogue elements not tied to any state. These weapons range from nuclear devices to chemical weapons to biological agents to computer viruses to long-range ballistic missiles. With the disintegration of the Soviet Union, opportunities for the thefts, and illegal sales and shipments, of such weapons have risen. In addition, numerous experts in the field of weapons production became "free agents," available for hire to the highest bidder.

Because such weapons could cause hundreds, thousands, maybe even millions of deaths—and economic disruption, even collapse— states will need to neutralize such threats swiftly and decisively. Conventional military options may prove unable to remove such a threat, especially if the master of the weapon in question is not at war with the potential victim. Second, the small size of such weapons makes their swift relocation easy in the face of a large and cumbersome conventional response. Thus the only way to remove the threat of a weapon of mass destruction may be to use a raid—quick, well-executed operation to prevent the relocation of the weapons. The small size of the raid makes the response appear proportional to the threat, and may prevent a broader conflict from erupting.

It is also worth mentioning that Western attitudes mitigate against taking (and inflicting) large numbers of casualties in military operations. The raid is a logical way to respond while minimizing the loss of human life.

The future development of nonlethal weapons will make the use of raids more attractive. Nonlethal weapons may some day be a factor in larger conventional operations, but the current research and development appears to favor preliminary use by smaller forces in environments where civilian casualties are likely, yet must be avoided, i.e., raids.

Opposition to nonlethal weapons, however, is emerging in some Third World countries. They fear that it will make intervention, and successful intervention at that, more likely. This attitude in some ways mirrors opposition by Third World states to Israel's 1976 raid in Uganda. Nevertheless, there remains at least one caveat. Nonlethal weapons may not prove as useful as their advocates suggest. Moreover, their deployment may signal to potential aggressors that their enemies are weak-willed. As such, nonlethal weapons may actually encourage some of the more thuggish types to undertake their depredations.

The likelihood that future unrest, whatever form it takes, will center around cities also suggests the increasing need for a "raiding" capability. The world is urbanizing, and future conflict will occur increasingly in these environments. Urban battlefields are horrendous, and combatants attempting to fight in such environs suffer exceedingly high casualties. Losses occur not only from weapons, but also the rapid spread of diseases. The heavy loss of civilian life and the concomitant destruction of the infrastructure will tax already inadequate methods of sanitation. Diseases are then free to run rampant through the combat units. The advantages many armies possess are lost in the concrete jungles. This is especially true of Western militaries. Their high-tech arsenals are degraded, and command, control, communications and intelligence become rudimentary. Forces native to the city have a distinct advantage. For examples of recent concrete battlefields consider Beirut, Mogadishu and Panama City, as well as Grozny and other Chechen cities.

Given the danger that the urban setting presents to conventional forces, one can see that raids will be one possible means of waging combat without committing one's main forces into an urban environment. Small, quick raids against a specific urban target would appear logical. Yet raids will be no panacea. When things go wrong, a raiding party will be at great risk. Consider the tragic loss of eighteen U.S. lives when the Rangers were trapped in Mogadishu while conducting a raid on October 3, 1993.

As countries around the globe continue to develop, their infrastructures become more vulnerable. Electric power grids, sources of energy (oil fields, nuclear and hydroelectric facilities), communications nodes, transportation hubs, water systems—all the elements that accompany modernization—are subject to attack. For countries unwilling to employ large conventional forces against an enemy state, raids against such vulnerable targets will appear tempting.

It has become fashionable for the current U.S. administration to fire cruise missiles at opposing countries, whether there is a military rationale or not. But not all modern infrastructures will be such easy targets for cruise missiles, or aircraft, in the future. First, countries will find more effective countermeasures to missile or aircraft raids, whether these be improved hardening of the target (including placing it underground) and/or developing stronger air defense systems. Second, there already exists a trend, evident in Iraq, to place potential targets in the heart of civilian communities. This will deter some cruise missile and aircraft attacks, for fear of inflicting civilian casualties. The only attack option available in some cases may be the commando-like raid.

Although I have focused so far more on raids outside of conventional war, let me assure the reader that there will remain a place for the raid in wartime. War will, sadly, continue to be a part of the human condition. With the end of the Cold War and the stability that the superpower rivalry ironically often provided, restraint on the international stage is disappearing. States will go to war over more minor matters and the very proliferation of states, as from the breakups of the Soviet Union and Yugoslavia, has already resulted in conflicts that could not have been foreseen a decade ago. And as the opportunities for war increase, the risks and penalties—as represented by potential superpower intervention—have decreased.

Thus it appears safe to say that raids, whether in a wartime context or a peacetime setting, will be with us in the future. The U.S. Army appears to recognize this since raids are mentioned in its manuals, both in wartime and in operations other than war (OOTW). For example, the recently revised FM 100-15 Corps Operations specifically mentions raids as one of its possible tasks in conducting operations other than war. It discusses Special Operations Forces (SOF) performing "their missions at the strategic, operational, or tactical level." In the section on OOTW, the manual lays out what actions raids may entail under the heading "Attacks and Raids."

Successful attacks or raids can create situations that permit seizing and maintaining the political initiative. Attacks and raids can also place considerable pressure on governments and groups supporting terrorism. Attacks and raids are planned and executed to achieve specific objectives other than gaining or holding terrain. Attacks and raids damage or destroy high-value targets or demonstrate the capability and resolve to achieve a favorable result. Raids are usually small-scale operations involving swift penetration of hostile territory to secure information, temporarily seize an objective, or destroy a target, followed by a rapid, preplanned withdrawal.

The section on attacks and raids in the manual mentioned above is followed by a discussion of antiterrorism, counterterrorism and, later, NEOs. And recognition of raids is readily found in other manuals. So the U.S. military does note the role that raids can play, although some may not find it to be sufficient. In fact, they may charge that the U.S. military is woefully unprepared for the future of warfare and "peacefare"—of which raids will be an increasingly important element.

Yet, I must conclude with a warning. Some will be tempted to suggest restructuring U.S. armed forces along the lines necessary for the conduct of raids or the prevention of raids by an enemy. The urge to radically reform will also be felt for the tangential or related issues, such as OOTW.

Inherent in this desire is a danger that we have witnessed in a slightly different context. Many defense experts want to see a lighter U.S. Army. In their minds, large, armor-heavy divisions ought to be assigned to the scrap heap of history. They argue that the future of conflict is centered around guerrilla, unconventional, irregular—whatever term you prefer—warfare. The 100-hour heavy armor blitz of the Gulf War, like Omdurman, was the last of its type, they say. Now it is time for the Army to get back to Rudyard Kipling's "savage wars of peace."

The "dump the heavy forces" argument makes mistakes that I hope we will not see with our growing focus on raiding, OOTW, and whatever other agendas come forward. For example, the light fighters see the debate all too often as an all-or-nothing disceptation—the Army as a whole should be geared to irregular warfare, rather than just elements of it. Then there is the confusion that light forces mean greater mobility. (To the battlefield, yes; once there, the heavy forces are more mobile.)

There is no more significant threat to the national security of the United States than the volatility and instability of Eurasia and its peripheries. The potential adversaries in these regions pose the greatest menace with their conventional, irregular, and possibly nuclear capabilities. The U.S. military must be prepared to fight all three types of threats. Raids will be just one weapon in the arsenal. The Israeli military is an example of this flexibility that is prepared to fight both armored forces and terrorists.

The future does not completely belong to irregular warfare. Some regional powers looking for quick victories will build conventional armies rather than follow the drawn-out, often unresolved, route of guerrilla tactics. Those states, peoples, or other groups that cannot initially compete at the conventional level will resort to irregular warfare. If one pares down one's forces to fight the guerrillas, they will in turn concentrate and fight a conventional war. Guerrilla war is a stage, a preparation for the ultimate conventional struggle. Remember that Saigon—after the U.S. withdrawal—was not captured by pajama-clad guerrillas but by armored forces. Therefore the existence of a guerrilla struggle is often clear evidence that an opposing conventional force is doing its job.

Guerrilla armies and irregulars do not change the geopolitical balance of power overnight. Armored columns, however, do. If the U.S. Army leans too far in one direction, I would rather it err on the side of being overly prepared for a quick, high intensity campaign and less prepared for a protracted, low intensity one, rather than the reverse.

U.S. heavy conventional forces give most nations on earth a reason not to contemplate sparring with America or her allies. This is all the more true after the Allied blitz through Iraq and Kuwait in 1991. As former U.S. Army Chief of Staff General Carl E. Vuono put it, "Conventional forces are also among the most effective tools for enhancing political stability in the international order." Yet, increasingly, one must realize that keen opponents are finding ways to negate the effects of conventional superiority. So rather than prepare only conventional options, the savvy military will have forces well trained to fight unconventionally, which will often mean using them in raids.

The 21st century will witness the continuation of the anarchic or multipolar state of international relations that prevails today. New threats may emerge that require declarations of war, the activation of conventional standing armies, the crushing of opponents' forces and the occupation of territory. Increasingly, however, new dangers will appear that do not respond to conventional threats and they will be removed only by the application of swift, focused raids by highly skilled and motivated troops. The actions of small groups of a nation's elite forces, during an anarchic peace, will no doubt echo the innovative courage and inspired leadership that have marked great raiders of every country and era.

Select Bibliography

DRAKE AT CÁDIZ

Bradford, Ernle. *Drake*. London: Hodder and Stoughton, 1965.

Hampden, John, ed. *Francis Drake: Privateer*. London: Eyre Methuen Ltd, 1972.

Martin, Colin, and Geoffrey Parker. *The Spanish Armada*. New York: Viking, 1988.

Mattingly, Garrett. *The Armada*. London: Jonathan Cape, 1959.

Sugden, John. *Sir Francis Drake*. London: Barrie & Jenkins, 1990.

THE FRENCH AND INDIANS AT DEERFIELD

Demos, John. *The Unredeemed Captive: A Family Story from Early America*. New York: Alfred A. Knopf, 1994.

Dillon, Richard H. *North American Indian Wars*. New York: Gallery Books, 1983.

Leach, Douglas Edward. *Arms for Empire: A Military History of the British Colonies in North America, 1607–1763*. New York: Macmillan, 1973.

Melvoin, Richard L. *New England Outpost: War and Society in Colonial Deerfield*. New York: W.W. Norton, 1989.

Williams, John. *The Redeemed Captive Returning to Zion*. Boston, 1706.

JOHN PAUL JONES OFF BRITAIN

Gilkerson, William. *The Ships of John Paul Jones*. Annapolis: Naval Institute Press, 1987.

Morrison, Samuel Eliot. *John Paul Jones: A Sailor's Biography*. Boston: Little, Brown, 1959.

Otis, James. *The Life of John Paul Jones*. Annapolis: Naval Institute Press, 1989.

Schaeper, Thomas J. *John Paul Jones and the Battle Off Flamborough Head: A Reconsideration*. New York: P. Long, 1989.

Seitz, Don C. *Paul Jones: His Exploits in English Seas During 1778–1780*. New York, 1917.

THE COSSACKS AT HAMBURG

Cazalas, E. *De Stralsund à Lunebourg: Episode de la Campagne de 1813*. Paris: L. Fournier, 1911.

Chandler, D. *Dictionary of the Napoleonic Wars*. New York: Macmillan, 1979.

Charras, Lt. Col. *Histoire de la Guerre de 1813 en Allemagne*. Paris, 1870.

Czernicheff. *Das Königreich Westphalen und seinen Armee im Jahre 1813*. Kassel, Germany: Heinrich Grobel, 1848.

Holleben, Gen. Maj. von. *Geschichte des Frühjahrsfeldzüges 1813 und Vorgeschichte*. Berlin: E. Siegfried Mittler & Sohn, 1904.

Nafizger, G.F. *Lutzen and Bautzen: Napoleon's 1813 Spring Campaign in Germany*. Chicago: Emperor's Press, 1992.

Plotho, C. *Der Kreig in Deutschland und Frankreich in den Jahren 1813 und 1814*. Berlin: Carl Friedrich, 1817.

Six, G. *Dictionnaire Biographique des Généraux & Amiraux Français de la Révolution et de l'Empire (1792–1814)*. Paris: Georges Saffroy, 1934.

Vaudoncourt, Gen. G. de. *Histoire Politique et Militaire du Prince Eugène Napoléon, Vice-roi d'Italie*. Paris: Librairie Universelle de P. Mongie, 1828.

MOSBY AT FAIRFAX COURT HOUSE

Boudry, Louis Napoleon. *Historic Records of the Fifth New York Cavalry*. Albany, NY: S.R. Gray, 1865.

Jones, Virgil Carrington. *Gray Ghosts and Rebel Raiders*. New York: Henry Holt, 1956.

Mosby, Col. John S. *Mosby's Memoirs*. Nashville: J.S. Sanders & Co., 1995 (reprint).

Scott, John (Major). *Partisan Life of Colonel John S. Mosby*. New York, 1867.

Williamson, James J. (Co "A"). *Mosby's Rangers: A Record of the Operations of the Forty-third Battalion of Virginia Cavalry*. New York, 1896.

MORGAN ACROSS THE OHIO

Bowman, John S., ed. *The Civil War Day by Day*. New York: Dorset Press, 1989.

Brown, Dee Alexander. *Morgan's Raiders*. New York: Konecky & Konecky, 1959, reprint 1993.

The Century Co., eds. *Battles and Leaders of the Civil War: The Tide Shifts* (Vol. III). Edison, NJ: Castle Books, 1883, reprint 1991.

Holland, Cecil Fletcher. *Morgan and His Raiders*. New York: Macmillan, 1942.

Keller, Allan. *Morgan's Raid*. Indianapolis: Bobbs-Merrill Co., 1961.

CUSTER AT THE WASHITA

Barnett, Louise. *Touched by Fire: The Life, Death and Mythic Afterlife of George Armstrong Custer*. New York: Henry Holt, 1996.

Connell, Evan S. *Son of the Morning Star*. New York: HarperCollins, 1984.

Dillon, Richard H. *North American Indian Wars*. New York: Gallery Books,

1983.

Stewart, Edgar I. *Custer's Luck*. Norman, OK: University of Oklahoma Press, 1955.

Utley, Robert M. *Cavalier in Buckskin: George Armstrong Custer and the Western Military Frontier*. Norman, OK: University of Oklahoma Press, 1988.

Welch, James, with Paul Stekler. *Killing Custer: The Battle of the Little Bighorn and the Fate of the Plains Indians*. New York: W.W. Norton, 1994.

Wert, Jeffrey D. *Custer: The Controversial Life of George Armstrong Custer*. New York: Simon & Schuster, 1996.

KOOS DE LA REY IN THE TRANSVAAL

de Wet, Christiaan Rudolf. *Three Years' War*. New York: Charles Scribner's Sons, 1902.

Farwell, Byron. *The Great Anglo-Boer War*. New York: Harper & Row, 1976.

Fisher, John. *The Afrikaners*. London: Cassell & Company, 1969.

Kruger, Rayne. *Good-bye Dolly Gray: The Story of the Boer War*. Philadelphia: J.B. Lippincott, 1960.

Le May, G.H.L. *British Supremacy in South Africa 1899–1907*. Oxford: Oxford University Press, 1965.

Pakenham, Thomas. *The Boer War*. New York: Random House, 1979.

Reitz, Deneys. *Boer Commando: An Afrikaner Journal of the Boer War*. New York: Sarpedon, 1993.

Warwick, Peter, Editor. *The South African War: The Anglo-Boer War, 1899–1902*. Harlow: Longman, 1980.

LAWRENCE OF ARABIA AT AQABA

Aldington, Richard. *Lawrence of Arabia*. London: Four Square, 1960.

Brown, Malcolm. *The Imperial War Museum Book of the First World War*. Norman: University of Oklahoma, 1993.

Fussel, Paul. *The Great War and Modern Memory*. Oxford: Oxford University Press, 1975.

Garnett, David, ed. *The Essential T.E. Lawrence*. Oxford: Oxford University Press, 1992.

Graves, Robert. *Lawrence and the Arabs*. New York: Paragon House, 1991.

GÜNTHER PRIEN AT SCAPA FLOW

Doenitz, Karl. *Deutsche Strategie zur See im zweiten Weltkrieg*. Munich: Bernard & Graefe, 1969.

Padfield, Peter. *War Beneath the Sea*. New York: John Wiley & Sons, 1995.

Rohwer, J., and G. Hummelchen. *Chronology of the War at Sea, 1939–1945*. Annapolis: Naval Institute Press, 1992.

Snyder, Gerald S. *The Royal Oak Disaster*. Novato, CA: Presidio Press, 1976.

Williamson, Gerald. *Aces of the Reich*. London: Arms and Armour, 1989.

BRITISH COMMANDOS AT ST. NAZAIRE

Darman, Peter. *Surprise Attack: Lightning Strikes of the World's Elite Forces.*
New York: Barnes and Noble Books, 1993.

Durnford-Slater, Brig. John. *Commando: Memoirs of a Fighting Commando in
World War Two.* Annapolis: Naval Institute Press, 1991.

Macksey, Kenneth. *Commando: Hit-and-Run Combat in World War II.* New
York: Jove Books, 1990.

McRaven, William H. *SPEC OPS: Case Studies in Special Operations Warfare.*
Novato, CA: Presidio Press, 1996.

Saunders, Hilary St. George. *The Green Beret: The Story of the Commandos
1940–1945.* London: M. Joseph, 1950.

THE CANADIANS AT DIEPPE

Atkin, Ronald. *Dieppe 1942.* London: Macmillan, 1980.

Brown, Anthony C. *Bodyguard of Lies.* New York: Harper & Row, 1975.

Saunders, Hilary St. George. *The Green Beret: The Story of the Commandos,
1940–1945.* London: M. Joseph, 1950.

Thompson, Reginald William. *Dieppe at Dawn.* London: Hutchinson, 1956.

Truscott, Lucien K., Jr. *Command Missions.* New York: Dutton, 1954.

FIRST CHINDIT

Fergusson, Bernard. *Beyond the Chindwin.* London: Leo Cooper/Combined
Books, 1995.

Hickey, Michael. *The Unforgettable Army: Slim's XIVth Army in Burma.* Kent,
UK: Spellmount Ltd, 1992.

Rolo, Charles J. *Wingate's Raiders.* New York: Viking, 1944.

Stibbe, Philip. *Return Via Rangoon.* London: Leo Cooper, 1995.

SECOND SCHWEINFURT

Bendiner, Elmer. *The Fall of Fortresses.* New York: G.P. Putnam's Sons, 1980.

Caidin, Martin. *Black Thursday.* New York: E.P. Dutton, 1960.

Galland, Adolf. *The First and the Last.* New York: Henry Holt, 1954.

Harvey, Maurice. *The Allied Bomber War, 1939–45.* Kent, UK: Spellmount
Ltd., 1992.

Newby, Leroy W. *Target Ploesti: View from a Bombsight.* Novato, CA:
Presidio Press, 1983.

Peaslee, Budd J. *Heritage of Valor: The Eighth Air Force in World War II.*
Philadelphia: Lippincott, 1964.

Wood, W. Raymond. *Or Go Down in Flame: A Navigator's Death Over
Schweinfurt.* New York: Sarpedon, 1993.

SKORZENY AT BUDAPEST

Infield, Glenn B. *Skorzeny: Hitler's Commando.* New York: St. Martin's Press,
1981.

Irving, David. *Hitler's War*. New York: Viking Press, 1977.

Keegan, John. *The Second World War*. New York: Viking Penguin, 1989.

Kursietis, Andris J. *The Hungarian Army and Its Leadership in World War II*. New York: Axis Europa, 1996.

Skorzeny, Otto. *My Commando Operations*. Atglen, PA: Schiffer Publishing, 1995.

Snyder, Louis L. *Hitler's Elite*. New York: Hippocrene Books, 1989.

Whiting, Charles. *Skorzeny*. New York: Ballantine Books, 1972.

BELGIAN PARATROOPERS IN THE CONGO

Reed, David. *111 Days in Stanleyville*. New York: Harper & Row, 1965.

Hoare, Michael. *Congo Mercenary*. London: Robert Hale & Co., 1974.

New York Times, The. November 17, 1961, p. 1.

Strobaugh, Donald R. *Journal—Operation Dragon Rouge, Stanleyville 24–27 November 1964*. Wiesbaden, Germany (unpublished), 1964.

Wagoner, Fred E. *Dragon Rouge: The Rescue of the Hostages in the Congo*. Washington, DC: National Defense University Research Directorate, 1980.

THE ISRAELIS AT GREEN ISLAND

Elder, Mike. *Flotilla 13*. Tel Aviv: Ma'ariv Book Guild, 1995.

Katz, Samuel M. *The Elite*. New York: Pocket Books, 1992.

———. *The Night Raiders*. New York: Pocket Books, 1997.

Melman, Yossi, and Dan Raviv. *The Imperfect Spies*. London: Sidgwick & Jackson, 1989.

Rabinovitch, Itamar. *The Boats of Cherbourg*. New York: Seaver Books/ Henry Holt, 1988.

THE AMERICANS AT SON TAY

Bowers, Frank R., B. Hugh Tovar, and Richard Schultz, eds. *Special Operations in US Strategy*. Washington, DC: National Defense University Press, 1984.

Brown, Ashley, and Jonathan Reed, eds. *The Unique Units*. (A volume in "The Elite" series.) London: Orbis Publishing, 1989.

David, Heather. *Operation: Rescue*. New York: Pinnacle Books, 1971.

Lipsman, Samuel, ed. *War in the Shadows*. (A volume in "The Vietnam Experience" series.) Boston: Boston Publishing, 1988.

Schemmer, Benjamin F. *The Raid*. New York: Harper and Row, 1976.

Vandenbrouke, Lucien S. *Perilous Options—Special Operations as an Instrument of U.S. Foreign Policy*. London: Oxford University Press, 1993.

Interviews: SGM (ret.) Joseph W. Lupyak; SGM (ret.) Galen C. Kittleson

DELTA FORCE AT DESERT ONE

Beckwith, Charlie A., and Donald Knox. *Delta Force*. San Diego: Harcourt Brace Jovanovich, 1983.

Kyle, James H. *The Guts to Try*. New York: Orion Books, 1990.

Ryan, Paul B. *The Iranian Rescue Mission: Why It Failed.* Annapolis: Naval
    Institute Press, 1985.

U.S. Department of Defense. "Rescue Mission Report" (typescript). August
    1980.

Valliere, John E. "Disaster at Desert One: Catalyst for Change," in *Parameters,*
    22 (Autumn 1982), 69–82.

ILLUSTRATION ACKNOWLEDGMENTS

The publishers are grateful to the following individuals and institutions for their
courtesy in making available the illustrations reprinted in this book. Every
effort has been made to obtain permission to use copyright material. The publishers apologize for any errors or omissions and would welcome these being
brought to their attention.

1, 3: Courtesy of The Public Record Office, London. 2: Courtesy of The
Prado Museum, Madrid. 4: Courtesy of The Schenectady County Historical
Society. 5: Courtesy of the U.S. Navy. 6: Courtesy of The Pierpont Morgan
Library, New York. 7, 8, 9: Courtesy of Emperors Press. 10, 11, 16: Courtesy
of the Library of Congress. 12: Warren A. Reeder. 13: Courtesy of the Filson
Club. 14: Courtesy of the University of Kentucky Library, Audio-Visual
Archive. 15: The Little Bighorn Battlefield National Monument. 17, 18, 20, 21,
24, 25, 30, 31, 32, 33: The Imperial War Museum, London. 19: The Cape
Archives, Cape Town. 22: The Bundesarchiv. 23: Courtesy of the Naval
Historical Center. 26, 27, 28, 29: Courtesy of the Department of National
Defence, Canadian Forces, Ottawa. 34: Courtesy of the U.S. Army Air Corps.
35: Herbert Kist. 36: USAISC. 39, 40, 41, 42, 43, 44: Courtesy of Donald R.
Strobaugh. 45, 46: Courtesy of the IDF. 47, 48, 49: Courtesy of the John F.
Kennedy Special Warfare Museum. 50, 51, 52, 53: Courtesy of James H. Kyle.

Index